Learning on the Blog

WILL RICHARDSON

Learning on the Blog

Collected Posts for Educators and Parents

CORWIN
A SAGE Company

A SAGE Company

FOR INFORMATION:

Corwin

A SAGE Company

2455 Teller Road

Thousand Oaks, California 91320

www.corwin.com

SAGE Ltd.

1 Oliver's Yard

55 City Road

London EC1Y 1SP

United Kingdom

SAGE Pvt. Ltd.

B 1/I 1 Mohan Cooperative Industrial Area

Mathura Road, New Delhi 110 044

India

SAGE Asia-Pacific Pte. Ltd.

33 Pekin Street #02-01

Far East Square

Singapore 048763

Acquisitions Editor: Hudson Perigo
Associate Editor: Allison Scott
Editorial Assistant: Lisa Whitney
Production Editor: Veronica Stapleton
Typesetter: Hurix Systems
Proofreader: Scott Oney
Indexer: Gloria Tierney
Cover Designer: Michael Dubowe
Permissions Editor: Adele Hutchinson

Printed in the United States of America.

Library of Congress Cataloging-in-Publication Data

Richardson, Will.

   Learning on the blog : collected posts for educators and parents / Will Richardson.

     p. cm.

   Includes index.

   ISBN 978-1-4129-9570-2 (pbk.)

1. Internet in education. 2. Teaching—Aids and devices. 3. Blogs. 4. Educational technology. 5. Educational change. I. Title.

   LB1044.87.R534 2012

   004.67'8071—dc23

             2011025627

This book is printed on acid-free paper.

11 12 13 14 15 10 9 8 7 6 5 4 3 2 1

Contents

Check out the blog at www.weblogg-ed.com!

About the Author

 A parent of two middle-school-aged children, **Will Richardson** has been writing about the intersection of social online learning networks and education for the past 10 years at Weblogg-ed.com and in numerous journals and magazines such as *Ed Leadership, Education Week,* and *English Journal.* Recently, he shifted his blogging emphasis to willrichardson.com. Formerly a public school educator for 22 years, he is a co-founder of Powerful Learning Practice (plpnetwork.com), a unique professional development program that has mentored over 3,000 teachers worldwide in the last three years. His first book, *Blogs, Wikis, Podcasts and Other Powerful Web Tools for Classrooms* (Corwin, 3rd Edition 2010) has sold over 80,000 copies and has impacted classroom practice around the world. His second book, *Personal Learning Networks: Using the Power of Connections to Transform Education,* was released in May, 2011. His articles have appeared in *Educational Leadership, EdWeek, English Journal, Edutopia,* and *Principal Leadership,* among others, and over the past six years, he has spoken to tens of thousands of educators in more than a dozen countries about the merits of learning networks for personal and professional growth. He is a national advisory board member of the George Lucas Education Foundation and a regular columnist for *District Administration* Magazine. Will lives in rural New Jersey with his wife, Wendy, and his children Tess and Tucker.

Introduction

Invitation to Participate in the Dialogue

Let's just start with this: there's a certain amount of irony in publishing a book of collected blog posts that can be freely found online. In fact, when the idea was first broached about a year ago, I have to admit I was a bit skeptical. What would be the point? But after thinking about it and testing the concept out with a number of educators as I traveled around last fall, I came around to the idea. Why?

A couple of reasons, really. First, regardless of the myriad ways to get information these days, people still like books, whether paper or digital. They like the idea that someone has taken the time and put in the effort to collect, consider, and synthesize ideas about the world and share them in a form that's easy to consume. It's true, as I make the case in some of the posts included here, that today we can write our own texts, compile our own curriculum if you will, and depend less on others to supply the information we need to learn. But it's also true that at least for the time being, many still value the idea that others are willing to do that for them. It's how they prefer to learn.

But second, and I think even more important, despite the best efforts of a growing community of educators espousing these ideas over the last decade, a large segment of the education population still has not been introduced to the profound shifts that social web technologies like blogs are undergoing as related to learning. Frankly, I've been amazed that now, even in 2011, I still meet large groups of people who, while they may have heard of Web 2.0 and may use Facebook or Twitter, still don't fully understand the implications of these tools for each of us as learners.

My hope is that this collection will meet some of those folks where they are, so to speak, and that it will frame a number of important questions and subsequent conversations that we need to be having about the changes we're witnessing in the world due to social media specifically as it applies to education. If a book of collected blog posts can do that, great . . . regardless of the irony.

Aside from having two children, blogging has no doubt been the most transformative thing that's happened to me in my life. When I first started weblogg-ed way (www.weblogg-ed.com) back in 2001, I never would have dreamed that it would turn out the way it has. It's connected me to people all over the world, people with whom I share a deep passion for education and technology and the potentials that go with them. My blogging has waned a bit the last couple of years for a variety of reasons, but it's still a place where I go on a regular basis to try to make sense of the world and to offer up mostly unfinished ideas to an audience that has been willing to tell me what it thinks almost 20,000 times over the years. It's been an amazing journey of learning, and I'm sincerely humbled by it on more levels than I can articulate.

The posts selected here, presented in their original form, represent some of the most discussed, most debated, and most fun pieces from the last six years or so. I tried to include some that still had relevance in the conversation today and some intended to make you think deeply about your own learning at this most interesting and challenging moment. For each group, I've added a short introduction to frame the larger themes and the salient questions. I hope you find them useful, and I hope these posts will lead to more questions and more conversations both online and off.

I want to thank my editor at Corwin, Hudson Perigo, for suggesting the idea and for her gentle prodding to move it toward completion. And I want to thank all of you who have read my blog over the years; you have made it what it is.

Finally, I've decided to donate all of my after-tax proceeds from the sale of this book to support technology initiatives for schools that need them the most. Once a year, I'll detail those efforts on my blog, the first installment of which will come in May of 2012. I've reaped more than enough rewards from these posts in their online form; hopefully, they'll help some deserving kids make their own forays into the blogosphere in meaningful ways.

Enjoy!

PART I

TEACHERS AS MASTER LEARNERS

On My Mind: Teachers as Master Learners

If we're to really understand the new learning opportunities for our students today, we ourselves have to be learners first. We have to build networks and communities online around the things we love so we can use them to teach and model for our kids how to do so on their own. Yet that's a difficult shift for many teachers to make. We think of these tools for instruction first, learning second. How can you become more selfish about learning? Can you give yourself the time to follow your own passions first? Is it hard to see yourself as a master learner? If so, why?

The roles and expectations of teachers are changing, but I wonder if we'll still be calling the adults in the room "teachers" in 20 years. The most important thing we can impart to our kids is a love of learning and the skills to learn well. Our expertise needs to be in connecting to other people, creating our own learning opportunities, and consulting with our students as to how they can achieve the same things autonomously. Maybe one day we'll be called "Master Learners" or something similar, because that's where our real value lies.

24 Feb 2010 08:45 am

As we continue to have conversations around change with the 800 or so practitioners we're **working with in PLP** (tinyurl.com/c48f92), I continue to be struck by the frustration I'm feeling at the seeming separation between teaching and learning. I know that this isn't new; **I've been writing about teachers' difficulties with being learners first here for a long time** (tinyurl.com/3vdmplj). When presented with the concept of building learning networks for themselves through the use of social learning tools, of making connections with other learners around the world who share their passions, many just cannot seem to break through the teacher lens and be "selfish" about it, to make it a personal shift before making a professional shift in the classroom. We want to teach with these tools first, many times at the expense, it seems, of making any real change in the way we see that learning interaction for our students because we don't experience that change for ourselves.

More and more, though, as I look at my own kids and try to make sense what's going to make them successful, I care less and less about a particular teacher's content expertise and more about whether that person is a master learner, one from whom Tess or Tucker can get the skills and literacies to make sense of learning in every context, new and old. What I want are master learners, not master teachers, learners who see my kids as *their apprentices **for learning***. Before public schooling, apprenticeship learning was the way kids were educated. They learned a trade or a skill from masters. When we moved to compulsory schooling, kids began to learn not from master doers so much as from master knowers, because we decided there were certain things that every child needed to know in order to be "educated." And we looked for adults who could impart that knowledge, who could teach it in ways that every child could learn it.

My sense is that we need to rethink the role of those adults once again, and that we're coming full circle. **George Siemens had a great post** (tinyurl .com/yknuclq) last week about "Teaching in Social and Technological Networks" and he asked the same question **that we had asked at Educon** (tinyurl .com/3ebcsr7): What is the role of the teacher? It changes:

> Simply: social and technological networks subvert the classroom-based role of the teacher. Networks thin classroom walls. Experts are no longer "out there" or "over there." Skype brings anyone, from anywhere, into a classroom. Students are not confined to interacting with only the ideas of a researcher or theorist. Instead, a student can interact directly with researchers through Twitter, blogs, Facebook, and listservs. The largely unitary voice of the traditional teacher is fragmented by the limitless conversation opportunities available in networks. When learners have control of the tools of conversation, they also control the conversations in which they choose to engage.

George goes on to suggest a totally different way of thinking about "teaching," one where "instead of controlling a classroom, a teacher now influences or shapes a network." And he discusses seven different roles that teachers will play, all of which are worth the read. The one that sticks out for me at least is the role of **modeling**, (bit.ly/mOyvfi) where he writes:

> Modeling has its roots in apprenticeship. Learning is a multi-faceted process, involving cognitive, social, and emotional dimensions. Knowledge is similarly multi-faceted, involving declarative, procedural, and academic dimensions. It is unreasonable to expect a class environment to capture the richness of these dimensions. Apprenticeship learning models are among the most effective in attending to the full breadth of learning. Apprenticeship is concerned with more than cognition and knowledge (to know about)—it also addresses the process of becoming a carpenter, plumber, or physician.

But I would argue it goes further than that, that apprenticeship for every student in our classrooms these days is not so much grounded in a trade or a profession as much as it is grounded in the process of becoming a learner. **Chris Lehmann** (tinyurl.com/yrz6yu) likes to say that we don't teach subjects, we teach kids. And I'll add to that: we teach kids to learn. We can't teach kids to learn unless we are learners ourselves, and our understanding of learning has to encompass the rich, passion-based interactions that take place in these social learning spaces online. Sure, I expect my daughter's science teacher to have some content expertise around science, no doubt. But more, I expect him to be able to show her how to learn more about science on her own, without him, to give her the mindset and the skills to create new science, not just know old science.

How we change that mindset in teachers is another story, however, and I know it has a lot to do with expectations, traditional definitions, outcomes, culture and a whole lot more. But we need to change it to more of what **Zac Chase** (tinyurl.com/3lnme6d) from **SLA** (tinyurl.com/3m56jfw) talks about in this snip I **Jinged** (tinyurl.com/ppnb3p) from the **"What is Educon?" video** (tinyurl .com/3kyce2z) posted by **Joseph Conroy** (tinyurl.com/3nqpe74). (Apologies for the audio and the stupid pop up ads.)

We still need to be teachers, but kids need to see us learning at every turn, using traditional methods of experimentation as well as social technologies that more and more are going to be their personal classrooms. How do we make more of that happen?

Sources: tinyurl.com/3ddbz7; tinyurl.com/ybejmzw

Personalizing Education for Teachers, Too

There's no question that the personalization or customization of education has begun. It's just not happening in schools yet. At a time when we have access to so much information and so many potential teachers or experts, we have to begin to think about how to give our students more of a voice in what they learn and how they learn it. In doing so, we can get to the goals and objectives that we set out for them in the curriculum. But the more accustomed our kids get to doing personalized learning outside of school, the more they'll demand it in the classroom. Can you model that for them?

28 Feb 2009 06:50 am

I finally got around to finishing up Sir Ken Robinson's new book **"The Element"** (tinyurl.com/3zhmwfr) which, for the most part, was a great read. He lays out a pretty compelling case for the power of passion in learning, and the absolute need for schools to help students identify their own passions through which they can learn just about anything they need. I've said in the past that the one thing I want from my own kids' teachers is for them to help them find what they love to do more than anything else and then support them in their learning endeavors around that topic. Unfortunately, that is not something the current public school system was built for.

Toward the end of the book, Sir Ken lays out the case for personalizing our kids' educations in the context of transforming (not reforming) schools:

> The key to this transformation is not to standardize education but to personalize it, to build achievement on discovering the individual talents of each child, to put students in an environment where they want to learn and where they can naturally discover their true passions (238). The curriculum should be personalized. Learning happens in the minds and souls of individuals–not in the databases of multiple-choice tests (248).

He argues that we should do away with the hierarchy of subjects and that we should work as hard as we can to customize, not standardize, each student's experience of schooling. Oh, to dream.

As I thought about those points, I started thinking about how we treat teachers and their learning as well. So much of professional development is throwing everyone in a room and having them learn the same stuff. Maybe there is some choice in the offerings, but by and large there is very little attempt at creating a customized professional development curriculum for teachers. Yes, we have our PIPs, but those usually address deficiencies or weaknesses, not passions.

The other day, I was having a conversation along these lines with a good friend who serves as the Director of Technology at a local school. We were talking about change, about how hard it is, and how long it takes. While he's done a great deal to move his school forward in terms of open source and social tools and technology in general, from a pedagogy standpoint, he had been racking his brain trying to figure out how to support individual teachers in these shifts. Finally, he came to the conclusion that the only way to do it was to create an individualized learning experience for each teacher, to take them where they are and mentor them, individually, to a different place. He's in the process of surveying each teacher to determine what technologies they currently use, what their comfort levels are, and what they are most passionate about. Then, using those results, he and one other tech educator at the school are going to start going one by one, talking about change, looking at tools, making connections, and shifting the pedagogy.

Whoa.

It echoes Sir Ken:

> Too many reform movements in education are designed to make education teacher-proof. The most successful systems in the world take the opposite view. They invest in teachers. The reason is that people succeed best when they have others who understand their talents, challenges, and abilities. This is why mentoring is such a helpful force in so many people's lives. Great teachers have always understood that their real role is not to teach subjects but to teach students (249).

Teachers are learners. If they're not, they shouldn't be teachers. In a world where we can engage in our passions through the affordances of connective technologies online, we need to be thinking about how to personalize the learning of the adults in the room as well as the kids. This is not the easy route, by any stretch, but it's the best route if we're serious about moving the education of our kids to a different place.

Source: tinyurl.com/d94mwc

Urgent: 21st Century Skills for Educators (and Others) First

These days, there is a lot of talk about students becoming fluent at "21st Century Skills" which, in all honesty, are skills that have been around for a long, long time. But at what point do we start demanding them of ourselves as teachers as well? Do we need to practice these skills regularly in order to better make sense of them for our students? I don't think there is any doubt. Publishing, participating, sharing ... these are all fundamental practices in the 21st-century world of learning. More and more, if you want your students (and others) to take you seriously, you're going to need to share your voice with the world.

Is there a better way to teach "21st Century Skills" than to practice them in your own learning? I can't think of one. Yet we're at an interesting moment where we're calling for all sorts of literacies for our students that very few of our 20th-century-trained and-educated teachers possess. What do we do about that? The first step is participation, because as you begin to share your ideas with the global community, you begin to understand what it means to create, think critically, collaborate, and more. Don't wait for the workshop. Dive in; the learning is fine.

09 Mar 2008 04:45 pm

Don't get me wrong, the **Thirteen Celebration conference** (tinyurl.com/ d3yysk) was a great event. Just getting the chance to hear Jane Goodall and Jean-Michel Cousteau speak about the amazing natural world and their sustained hope that we can undo much of the damage we've done to it was in and of itself worth it. That and being able to listen to the passion of Diane Ravitch and Deb Meier as they discussed the ills of education gave me more than enough to think about.

But here's the thing that's giving me the most angst. (Hey, I haven't been too angsty in a while, have I?) For all of the experts and scholars and pundits who were staking out a part of the conversation about educational reform, I couldn't help leaving there wondering how many of them really have a sense of the changes that are afoot here. I looked up a whole bunch of the names of the presenters and I could only find a handful that have any real Read/Write Web footprint that would allow me to consider them to be a part of my network. And worse, it was painfully obvious by their death by PowerPoint presentation styles that their own adoption of technology as a communication tool not to mention a networked learning tool left a great deal to be desired. The governors, the state superintendents, the consultants . . . from none of them did I get the sense that they could give a great response to a request to model their uses of technology to teach and learn effectively, especially in the context of networks.

All of which raise a number of questions:

First, am I a snob? Out to lunch? I mean it. I feel like it sometimes when I go to an education conference with 6,000 attendees and virtually no Internet access where almost no one who is presenting is modeling anything close to great pedagogy with technology. (That doesn't mean, btw, that they are not great teachers or thinkers.) Where just about the only technologies represented on the vendor floor deal with assessment or classroom displays. I mean, I know I'm a one-trick pony in terms of what my frame of reference is (so no need to remind me again), but shouldn't I be at least getting some sense that the people who are making the decisions understand on some level what we here are jammering about every day, the transformation that's occurring, the amazing potentials of this? I feel like I have to be missing something here, that it must be me.

Which leads to the second question which is how in god's name can we talk seriously about 21st Century skills for kids if we're not talking 21st Century skills for educators first? The more I listened, the less I heard in terms of how we make the teaching profession as a whole even capable of teaching these "skills" to kids. Sure, there were mentions of upgrading teacher preparation programs and giving teachers additional time in the school day to collaborate, etc. But the URGENCY was all around the kids. Shouldn't the URGENCY be all about the teachers right now?

Finally, I was struck by how difficult it felt to accept much of what I was hearing because, and this is something that is really concerning me (seriously), few if any of these folks had the network creds to be "trusted." Now I know this is an admission that is going to get me in trouble, and it likely should. But it is also a consequence of being rooted so deeply in this network. It's not that I distrust their "traditional" creds out of hand, but it's almost like for me, these days, if you're not doing at least a little bit of social, networked learning and publishing that I can tap into and track and engage with, I'm just not as inclined to buy in when you're talking about reforming education with or without technology.

Which leads to the following conclusions. First, (and this really has little to do with the larger point of this post but is stuck in my head) if you want to do one thing to save the world, become a vegetarian, today. Right now. As Cousteau said, any time you eat a steak or a chicken, you are cutting down a tree in the Amazon. Beef cattle graze, use up the land, then they plant soybeans to feed the chickens being farmed all over the world. Rainforests gone. Our carnivorousness is killing the planet. That means no pot roast nachos either. End of sermon.

And, second, if you want your ideas to resonate with me and to be taken seriously, don't just talk. Engage. Publish. Converse. Add your voice to the network of people who are living these ideas every day. I'll use **Mr. Stager** (bit.ly/oDuGhV) as an example here, since I know he'll most likely have something to say on the topic. I'd heard of Gary before he started blogging, but the fact that he's now willing to put his ideas out here and invest in the network, whether I agree with him or not, garners my respect and makes me more open to his ideas. I can think of a number of folks in this arena who I can't say that about.

Rant over. Be gentle . . .

Source: tinyurl.com/3xbk8d

Why Is It So Hard for Educators to Focus on Their Own Learning?

How do you learn? How has your learning changed? What new ways of learning can we take advantage of? In these times of high-speed change, it's doubly important for educators to be reflective about the way they learn, not just the way they teach. When you're introduced to a new tool, get in the habit of asking, "How can I use this in my own learning practice?" not "How can I use this in the classroom?"

17 Jul 2007 03:10 pm

That's a question that I'm really trying to get my brain around of late. In the past few weeks, I have really ramped up my rhetoric to teachers in terms of trying to get them to examine how these technologies challenge their own personal learning. How can the connections we make with these tools affect their own learning practice? How can they begin to understand what the implications for learning are for their students until they at some level understand them for themselves? And so on. And for the most part, heads nod politely in agreement.

But, here's the thing. By and large, most of the questions that come up during the workshop or the presentation run along the lines of "how do we keep our kids safe with this stuff?" or "if I want to put up my homework for my kids is it better to use a blog or a wiki?" or "so parents could subscribe to these RSS feeds, right?" All good, useful, legitimate questions. But very far removed from the personal learning focus I've been trying to articulate. In fact, when I stand by these teachers and hear their questions, when I look at them directly and say "well, that's a great question, but I really want you to focus on your own practice here, your own learning," more often than not what I get is a scrunched up face, a biting of the lower lip, a feeling that their brains are saying "AAARRRGGGHHH."

And even as I sit in this session with Tim Tyson at Building Learning Communities, one principal says "I want to learn more about these tools so I can help my teachers use them in the classroom." I want to jump up and say "No! You are missing a step! You want to learn more about these tools for yourself so you can help your teachers learn from them too."

So what's that all about? Is it just habit? Is it just such a focus on curriculum delivery that "learning" is all about how to do that job better? Is changing the way we do our own business just too darn hard? Or is this such a huge shift, this idea that we can actually learn through the use of technology that most people just don't think they have to go there, that they can just keep using it as a way to communicate without the surrounding connective tissue where the real learning takes place?

Or, maybe it's just me . . .

Source: tinyurl.com/2cycbk

Teaching Ourselves Right Out of a Job

This post from six years prior to this writing still resonates deeply with me. I'm more convinced than ever that our most successful students will be those who can teach themselves, those who can create their own classrooms and communities and curricula around the things they want or need to learn. To that extent, we need to rethink much of our role—as the traditional teacher, the content expert, the classroom manager—in favor of model learner and connector of students to other teachers outside the classroom walls. In what ways can you begin doing this with your students?

14 Sep 2005 05:00 am

We had an interesting conversation at dinner last night revolving around the changes that are occurring in classrooms these days. Since we're in the middle of our Tablet PC pilot at our school right now, I know this is especially acute as I've seen some pretty remarkable things this first week with teachers and students. But last night we were talking about the access to information that many (but not all) students and teachers have via the web. And we were talking about how few educators had made the Internet a significant part of their practice. If we're entering a world where much of what we do in business, communication, politics, etc. will be done online, we have to prepare our students for that reality. And the most effective way to teach these skills is to master them ourselves.

Case in point: I was talking with a math teacher who is a part of our pilot, and he told me that in the course of his lesson on Monday he used a term that was unfamiliar to his students. Rather than simply give them a definition, he modeled his own practice by having his students watch as he went from the OneNote page he was projecting via his tablet, opened up a browser, surfed over to **Wikipedia** (tinyurl.com/268zt), looked up the definition, and started a discussion about not only the math but about the workings of the site. Now I would bet that only a handful of teachers would model that same process.

And why is that? I'm back to that again, I know. The web and these technologies have transformed the way I learn, provided me with many teachers who push my thinking, given me the potential to direct my own education as it is. Why don't more educators make it a part of their own practice?

What I realized more clearly last night is that for many teachers, the idea of teaching kids to be able to access information and find mentors and communities of practice basically means teaching themselves out of their jobs, at least as they know it. I mean, at some point, we're going to have to let go of the idea that we are the most knowledgable content experts available to our students. We used to be, when really all our students had access to was the textbook and the teacher's brain. But today, we're not. Not by a long stretch. And we don't need to be. What we need to be is connectors who can teach our kids how to connect to information and to sources, how to use that information effectively, and how to manage and build upon the learning that comes with it. That's a much different role than "science teacher" or "math teacher." Now I'm not saying that subject matter expertise is irrelevant and that there aren't core concepts that discipline specific teachers shouldn't teach. But they should be taught it a much wider context, not in the fishbowl that is our traditional classroom.

This is a scary idea, I think. But it takes me back to something I wrote a couple of days ago that was almost a throw away line at the time but one that got me thinking much more deeply about all of this stuff:

The best teachers are the ones we find, not the ones we're given.

There's much more to write about that . . .

Source: tinyurl.com/5j87vx

The Next Generation of Teachers

One of the biggest challenges for teachers right now is reframing the way they think about access and technology in the classroom. For instance, how can we begin to think of the devices we give to kids as connections, not just computers? How can we aspire to create connections outside of our classrooms instead of keeping everything within the four walls? It's easy to talk about "why not." How instead do we keep the conversation focused on the potentials rather than the problems?

21 Mar 2007 11:24 am

Last night I got the chance to spend a couple of hours with about 20 graduate students in education at a pretty large school here in New Jersey during a class they are taking in educational technologies. We ended up talking about a lot of the shifts that are occurring right now, the tools, and the challenges that face them as they enter the profession. Now, I was really impressed with the level of the conversation and the sincere questions that they shared. But I was also struck by how much of a reality check it was for me, at least.

The general sense from the group was "yeah, but" once again. Yeah, but we have these kids who are going to abuse these technologies if we open them up. Yeah, but we're going to be out there on our own if we decide to use these technologies. Yeah, but I don't have enough time to make this a part of my own practice. Yeah, but, etc. (And please, if any of those in attendance are reading this, feel free to chime in.) At one point I said something along the lines of "you know, there's a lot of pressure on you in my circles because many people think nothing is going to change until the old guard retires out and you guys take over." Well, that didn't float very well. I got the sense that most didn't want to accept that challenge or felt it was just too daunting. And at another point, after going through a list of reasons why using these ideas were going to be difficult, I said "yes, but you know there is nothing stopping you from changing the way you learn." Not sure how well that went over, either.

I don't mean to come across as disparaging to any of these students. You could tell they were by and large smart and sincerely interested in the discussion. But I guess I was hoping for more, though I'm also not entirely surprised I didn't get it.

One other thing. A couple of them noted that this one class (which was an elective, by the way) was the only (stress . . . ONLY) time in their grad program that they had talked about technology in a pedagogical sense.

As much as I want it to be otherwise, the reality here is that we're just not getting it done on so many levels.

Source: tinyurl.com/35ujzp

Teachers as Learners Part 27

The problem with teachers when they become teachers is that their definition of what it means to be a teacher is based on the teachers they had growing up. (Say that three times fast.) Why is this a problem? Because it leaves few other lenses through which to surmise what a teacher is. I wonder how many would go into teaching if the definition were changed to "one who learns with her students."

30 Aug 2006 07:29 pm

The whole integrating technology discussion that many have been chronicling of late has been sticking in my craw for a couple of reasons. First, a couple of weeks ago I had a bad teacher day while I was doing some training, the kind that really gets me pessimistic about how difficult a road this is going to be.

With this particular group, it was made clear that the only reason they were in attendance was that they were getting paid for the day, that any teacher who came in during the summer and wasn't getting paid was ruining it for everyone else, that the technology wouldn't work in their classrooms anyway, that they didn't have time to practice what they were learning, that, well, fill in the blank. It was one of those days, and they don't occur very often, but it was one of those days when I walked out of the room thinking "Thank god my kids don't go to this school."

Depressing, to say the least.

The second reason is that it's becoming exceedingly clear that we have an outdated perception of what teachers need to be. Like David, more and more I think there is a "T" word that we should stop using, only mine isn't technology. It's teaching. And let me say up front that this is one of those "I'm blogging this so people will help me figure out what it is I think" posts as my thoughts are still somewhat murky. But here goes.

When we say "teacher," what we are really saying is "the person in the classroom to whom students look for knowledge" or something like that. In the traditional classroom that almost all of us grew up in, the teacher was the focal point, the decision maker, the director, the assessor. Teachers, well, teach, or try to. We hire teachers based on how well they know their subject matter and how well we think they can deliver it to students. Teaching, the way most of us see it, is all about imparting knowledge in a planned, controlled way.

In a world where knowledge is scarce (and I know I'm using that phrase an awful lot these days), I can see why we needed teachers to be, well, teachers. But here's what I'm wondering: in a world where knowledge is abundant, is that still the case? In a world where, if we have access, we can find what we need to know, doesn't a teacher's role fundamentally change? Isn't it more important that the adults we put into the rooms with our kids be *learners* first? Real, *continual learners*? Real models for the practice of learning? People who make learning transparent and really become a part of the community?

I hesitate to make blanket statements about teachers because a) they are seldom appropriate (the statements, that is) and b) they get me in trouble. But when I ask myself what percentage of the thousands of teachers I've worked with over the past two years are practicing *learners*, I have a hard time convincing myself that it's more than half. Maybe even one-third.

I'm not saying this is necessarily their fault. We teach teachers to teach, we don't teach teachers to learn. Even in professional development, we teach them stuff they need to be better teachers, but do we give them the skills they need to be better learners? Do we evaluate them on what they've been reading? On what they've been writing? On their reflectiveness?

There is a section in **Henry Jenkins' book** (tinyurl.com/3p9wxv2) that somewhat goes to this titled "Collective Intelligence and the Expert Paradigm." I'm going to blog about it in this context when I next get a chance (which might not be for a few days).

But for now, I'll keep trying to think it through. What if we hired learners first?

Source: <u>tinyurl.com/69xkx6t</u>

Unlearning Teaching

So let's push this reenvisioning of teaching even further. What if teachers and students were co-learners, co-creators in the process? What if we saw our classrooms as laboratories and our own roles as participants in the work instead of leaders of the work? What if we entered our classrooms with a sense of "not knowing" and shared that lens with our students? For some, this is a terrifying thought. But at a moment when knowledge is simultaneously ubiquitous and evolving, I'm not sure we have much choice.

18 Aug 2010 07:04 am

> Rather than teachers delivering an information product to be "consumed" and fed back by the student, co-creating value would see the teacher and student mutually involved in assembling and dissembling cultural products. As co-creators, both would add value to the capacity building work being done through the invitation to "meddle" and to make errors. The teacher is in there experimenting and learning from the instructive complications of her errors alongside her students, rather than moving from desk to desk or chat room to chat room, watching over her flock.

I love this vision of teaching from Erica McWilliam, articulated in her 2007 piece "**Unlearning How to Teach**" (tinyurl.com/27w8h4w) (via my **Diigo** [tinyurl.com/g5uja] network). I know the idea isn't new in these parts, but the way she frames it really resonates. And it speaks to some important aspects of network literacy and the teacher's role in the formation of and the participation in those student networks. At the end of the day, as she suggests in the quote above, we have to add value to the process, not simply facilitate it. Here's another snip that gets to that:

> A further point here—if we consider the student's learning network as a type of value network, then, we must also accept that such a network allows quick disconnection from nodes where value is not added, and quick connections with new nodes that promise added value—networks allow individuals to "go round" or elude a point of exchange where supply chains do not. In blunt terms, this means that the teacher who does not add value to a learning network can—and will—be by-passed.

I think that's one of the hardest shifts in thinking for teachers to make, the idea that they are no longer central to student learning simply because they are in the room. When learning value can be found in a billion different places, the teacher has to see herself as one of many nodes of learning, and she has to be willing to help students find, vet, and interact with those other nodes in ways that place value at the center of the interaction, meaning both ways. It's not just enough to add those who bring value; we must create value in our networks as well.

Another interesting point in the essay suggests that because of our emphasis on knowledge in the schooling process, we are actually creating a more ignorant society. I greatly admire **Charles Leadbeater's** (tinyurl.com/6frh6p) work (If you haven't read **"Learning from the Extremes"** [tinyurl.com/3xlcnfr] (pdf) you need to), and this somewhat extended quote really got me thinking:

In a script-less and fluid social world, "being knowledgeable" in some discipline or area of enterprise is much less useful than it was in times gone by. In *The Weightless Society* (2000), Charles Leadbeater explains the reason for this by exploding the myth that we are becoming a more and more knowledgeable society with each new generation. Leadbeater's view is that we have never been more ignorant. He reminds us that we have a much less intimate knowledge of the technologies that we use every day than our forebears had, and will continue to experience a growing gap between what we know and what knowledge is embedded in our manufactured environment. In simple terms, we are much more ignorant in relative terms than our predecessors.

But Leadbeater makes a further point about our increasing relative ignorance that is highly significant for teaching and learning. It is that we can and must put this ignorance to work—to make it useful—to provide opportunities for ourselves and others to live innovative and creative lives. "What holds people back from taking risks," he asserts, "is often as not . . . their knowledge, not their ignorance" (p. 4). Useful ignorance, then, becomes a space of pedagogical possibility rather than a base that needs to be covered. "Not knowing" needs to be put to work without shame or bluster . . . Our highest educational achievers may well be aligned with their teachers in knowing what to do if and when they have the script. But as indicated earlier, this sort of certain and tidy knowing is out of alignment with a script-less and fluid social world. Out best learners will be those who can make "not knowing" useful, who do not need the blueprint, the template, the map, to make a new kind of sense. This is one new disposition that academics as teachers need to acquire fast—the disposition to be usefully ignorant.

As a parent, and I know I keep coming back to this lens more and more these days, I want my kids and their teachers to be "usefully ignorant." It's the basis of inquiry, and that type of learning can't happen unless we give up this notion that we can "know" the answer and that it can be tested in a neat little short answer package. The world truly is "script-less," and the more my kids are able to flourish with "not knowing" the more successful they will be. Just that concept will require a lot of "unlearning" when it comes to teaching and schools in general.

So how are you unlearning teaching?

Source: tinyurl.com/22rwjso

"What Did You Create Today?"

It's hard for me to separate my role as a parent from my role as an educator, and this post captures much of my current frustration. Our conversations about schools rarely center on anything more than grades and homework. Still, I know this is mostly due to the expectations of the system, but I still believe that we can create meaningful, relevant opportunities for students to contribute to the world while developing passion for and adhering to all of those traditional expectations. How we make that engagement happen in our classrooms is, I think, the most important inquiry for an educator.

22 Aug 2009 08:55 am

In a couple of weeks, both Tess and Tucker will be starting their first day at brand new schools. They'll know no one, have all new teachers, new surroundings, and, hopefully, new opportunities. I'm not sure they're totally at peace with these changes, but as I keep telling them, it's the kind of stuff that builds character. (I keep regaling them with school switching stories of my own, the most challenging being when my mom moved us out to New Jersey from Chicago when I was beginning 6th grade and three days before school started I was wading barefoot in a creek, stepped on a broken bottle, and ended up with 10 stitches in the bottom of my foot and a pair of crutches for the first week of classes. Talk about character building.) Wendy and I have been trying to prepare them for this shift as best we can, and while I know it's a bit scary for them, I'm really hopeful the change will be good for them on a lot of different levels.

What I'm most hopeful for, however, is that their stories about school will change. Last year, far too much of the reporting about their days started with "I got a ___ on my ___ test!" or "Yes, I've got homework" (said in the same voice as one might say "Yes, I've got ringworm"). School was something that rarely sparked a conversation about learning. Usually, it was a topic to be avoided or ignored. I hope to hear more excitement this year, more passion about learning, more thinking and doing. To that end, I've been coming up with a mental list of the types of questions I'm hoping they might answer:

- What did you make today that was meaningful?
- What did you learn about the world?
- Who are you working with?
- What surprised you?
- What did your teachers make with you?
- What did you teach others?
- What unanswered questions are you struggling with?
- How did you change the world in some small (or big) way?
- What's something your teachers learned today?
- What did you share with the world?

- What do you want to know more about?
- What did you love about today?
- What made you laugh?

I think their answers to those questions (and others that I'm hoping you might add below) would tell me more about what they learned than any test or quiz or worksheet that they brought home for me to sign. And here's the deal; I expect them to be talking answers to these types of questions every day. As a parent, I think I have every right to expect that my kids are immersed in spaces where learning is loved and enjoyed and shared every single day. Classrooms where they are engaged in meaningful work that makes them think, a majority of time doing stuff that can't be measured by some impersonal state test. (I can give them software to do much of that.) Where the adults that surround them are models for that learning work themselves. Is that too much to ask?

New schools, new opportunities, renewed expectations. We'll see how it goes . . .

Source: tinyurl.com/medywy

Get. Off. Paper.

Becoming fully functional in these online spaces will require a letting go of some of our old learning practices and habits. And probably the most difficult for many is to give up paper and pen. Here's the reality as of February 2011: 98% of all knowledge is digitized, Amazon is selling more e-books than paperbacks, and being able to navigate in digital multimedia spaces is now a literacy according to most. For those of you still wedded to paper, I know this is hard. But do you really think the kids in your classrooms will ever take notes with paper and pen in their lives? Digital is a different, I think better, beast. It has to be a seamless part of our learning lives.

13 Nov 2008 05:35 pm

The other day I was talking to a school administrator about an upcoming hands-on workshop and she asked if I could e-mail her the schedule to hand out the morning of the event. For some strange reason I just said "Nope. No paper."

After a short silence, she said, "Oh . . . ok."

"No, I mean it," I said. "We're going to be spending the whole day online; there is no reason to bring paper."

"Really?"

"Really."

"No paper," she said, thinking, finally adding "How exciting!"

Now I don't know that I've ever thought of no paper as exciting, necessarily, but I continue to find myself more and more eschewing paper of just about any kind in my life. My newspaper/magazine intake is down to nearly zero, every note I take is stored somewhere in the cloud via my computer or iPhone, I rarely write checks, pay paper bills or even carry cash money any longer, and I swear I could live without a printer except for the times when someone demands a signed copy of something or other. (Admittedly, I still read lots of paper books, but I'm working on that.) Yet just about everywhere I go where groups of educators are in the room, paper abounds. Notebooks, legal pads, sticky notes, index cards . . . it's everywhere. We are, as Alan November so often says, "paper trained," and the worst part is it shows no signs of abating.

At one planning session I was in a few weeks ago, twenty people were all furiously scribbling down notes on their pads, filling page after page after page. The same notes, 20 times. (I'd love to know where those notes are now.) At the end of the session, I gave everyone a TinyUrl to a wiki page where I had stowed my observations and asked them to come in and add anything I missed. Two people have.

At the end of a presentation a few days ago with a couple of hundred pen and paper note taking attendees (and the odd laptop user sprinkled here and there) I answered a question about "What do we do now?" by saying "Well, first off, it's a shame that the collective experience of the people in this room is about to walk off in two hundred different directions without any way to share and reflect on the thinking they've been doing all day. Next year, no paper."

I don't think most were excited. It all reminds me of the time last year when I got to an event and the person in charge had copied, collated, stapled and distributed six paper pages that she had printed of my link-filled wiki online to 50 or so participants.

"It's a wiki," I said. "You can't click the links on paper!"

"I know," she replied. "I just need to have paper."

Um, no. You don't.

Does anyone think most of the kids in our classes are going to be printing a bunch of paper in their grown up worlds? If you do, fine; keep servicing the Xerox machine. But if you don't, which I hope is most of you, are you doing as much as you can to get off paper?

Source: tinyurl.com/66jbmj

Opportunity, Not Threat

Parents are an important piece in this puzzle we're trying to put together around change. It's difficult for most parents to envision their children experiencing school differently from the way they did themselves. Yet we need to help them see just that. At the same time, can we as educators really wait for parental approval to create a different learning path for our students? As the "experts" in education in our local communities, do we have a responsibility not only to change what we do in the classroom but also to bring parents into the process?

07 Apr 2010 10:40 am

Let's just start with this money quote from Michael Feldstein in a comment on the must read post by Jim Groom titled "Networked Study":

It's hard to change the culture of education without getting the kids before their thinking processes begin to ossify, but in order to do that, you have to contend with their parents who, however well-intended, didn't have the benefit of the kind of education you're trying to provide their kids and often see it as more of a threat than an opportunity.

To me, that's the most interesting piece of this conversation right now, how to move the parents' perspective of the nascent, non-traditional models of education to one that really embraces the opportunities that online communities and networks are creating for meaningful learning. I know that when I talk about my aspirations for my own kids, and I start going down the road that the traditional college degree is only one of many options for them, that they may be able to cobble together a more meaningful education (depending on what they want to do) through travel and apprenticeships and self-directed experiences and not end up in mountains of debt, most respond with all sorts of reasons why not going to college is a risk, "especially in this job market." (As if college grads are stepping into great jobs these days anyway.)

Here's another quote that speaks to this idea, this time from Anya Kamenetz's new book DIY U: Edupunks, Edupreneurs, and the Coming Transformation of Higher Education:

I've had a number of parents tell me that as much as they truly believe the educational landscape is changing, it's hard for them to sanction their own kids being a part of that change. "To some degree I lack the courage of my convictions . . . I'm developing very strong convictions that the existing system is fundamentally and probably irreparably broken, but I would not yet take my kids out of their school," Albert Wenger at Union Square Ventures said. "It's one thing to experiment by investing money in start-ups or reading books, and it's another to experiment with your own children."

There are so many levels to this from a parenting perspective that it's hard to know where to begin. Most parents think their kids' schools are doing just fine based on the assessment systems we currently have in place. Most parents see the traditional track from high school to college as success. Most parents are ok with "online courses" and can use them to check the technology box since they don't radically disrupt the status quo. Most parents have no clue as to what that change they might be sensing really looks like. They don't, as Jim Groom writes, see education as "the biggest sham going."

Whoa.

The roll your own education "movement" is obviously not just a disruption to parents; it's a threat to educators as well. The question of how to help them find opportunity here is one we'll be struggling with for decades, no doubt.

But isn't the bottom line here helping our kids take advantage of the opportunities? This comment by Michael Feldstein about how kids don't have the ability to direct their own learning echoes the ridiculous expectations floated by Mark Bauerline in the Dumbest Generation, that somehow, these kids today are supposed to learn this all on their own:

> It's not like student-centered education was created by the edupunks. And yet, students fail to learn in these classes all the time. The high drop-out rate in community colleges reflects a lot of different factors, but one major one is surely that many students who go there do not have the skills to take charge of their own education, no matter how much you try to empower them. I have not been given reason to believe that the digital version of this approach will be wildly more successful than the analog version.

Is it any wonder they can't "take charge of their own education" when that self-directed love of learning on their own was driven out of them by second grade, when no one has ever allowed them or taught them how to do that? And are we at the point where we can begin to give them reasons to believe? Are we? (In fairness, Feldstein accedes to this later in the thread.)

The irony here is obvious: right now, as it's currently structured, traditional schooling is in many ways the threat, not the opportunity that many still see it as. How we make that message digestible to parents is, I think, the most interesting question of all. And how we do it in ways that don't drive people to the edges but instead help them work in the messy middle and make sure we ultimately keep in mind what's best for the care of our kids is the most challenging part of all. To that end, I love this quote from a recent must read Mark Pesce post:

> There is no authority anywhere. Either we do this ourselves, or it will not happen. We have to look to ourselves, build the networks between ourselves, reach out and connect from ourselves, if we expect to be able to resist a culture which wants to turn the entire human world into candy.

This is not going to be easy; if it were, it would have happened by itself. Nor is it instantaneous. Nothing like this happens overnight. Furthermore, it requires great persistence. In the ideal situation, it begins at birth and continues on seamlessly until death. In that sense, this connected educational field mirrors and is a reflection of our human social networks, the ones we form from our first moments of awareness. But unlike that more ad-hoc network, this one has a specific intent: to bring the child into knowledge.

Yep.

Source: tinyurl.com/6bf6k2a

Response to Jay Matthews
at the Washington Post

Do we require new skills and literacies to navigate this new world or do we just have to get better at the things we've been doing all along? That's an important question to consider as we think about learning in online networks. I think they encompass new and different requirements that bear little resemblance to what most would currently describe as literacy. How do we help students navigate the world as they are experiencing it instead of the way we experienced it?

05 Jan 2009 09:23 am

Jay Matthews wrote a piece in the Post this morning titled "The Latest Doomed Pedagogical Fad: 21st Century Skills" to which I replied what follows. Would be interested to hear your thoughts, here or there . . .

I don't disagree that the majority of "21st Century Skills" are nothing new, and that we should have been teaching them all along. As computer and online technologies evolve, we have more tools that we can use to teach those skills in perhaps more relevant or compelling ways. But that depends on the teacher's familiarity and comfort level with those technologies, obviously.

What is different here, though, is something that is not being articulated by the Partnership or many others, and that is the learning that can be done (and is being done already) using online social tools and networks. I'd point you to a recent MacArthur Foundation study which concludes that "New media forms have altered how youth socialize and learn" and that this has very important implications for schools and teaching (tinyurl.com/55a878, pdf). While most kids' uses of these technologies are "friendship based," the more compelling shift is when their use is "interest based" or when they connect with other kids or adults around the topics or ideas they are passionate to learn about. With access to the Internet, and with an understanding of how to create and navigate these online, social learning spaces, opportunities for learning widely and deeply reside in the connections that we make with other people who can teach or mentor us and/or collaborate with us in the learning process.

That, I think, is where we find 21st Century skills that are different and important. Sure, those connections require a well developed reading and writing literacy, and critical thinking and creativity and many of the others are skills inherent to the process. But this new potential to learn easily and deeply in environments that are not bounded by physical space or scheduled time constraints requires us as educators to take a hard look at how we are helping our students realize the potentials of those opportunities.

Having blogged now for seven years and having learned in these interest or passion-based online networks and communities for almost as long, it's hard to begin to describe how different it is from the classroom teaching that I did for 18 years in a public high school. My learning is self-directed, and everyone in these virtual classrooms wants to be there because they too are interested in pursuing their interests. They come from all over the world, all different cultures, all different experiences, a diversity that is hard to fashion in most school classrooms. We share our learning openly, admit anyone into the conversation, and constantly seek to make each other smarter. But while that can sound like a pretty positive and powerful space, it is fraught with complexity. We have to learn to read not only texts but to edit them as well, not just for accuracy but for bias, agenda and motive. In the online learning world, we have to be full fledged editors, not just readers, because the traditional editors are gone from the process. And, we have to be creators as well. In order for us to be found by potential teachers and collaborators, we need to have a presence, a footprint. I'm fully convinced that my own kids need to publish, need to establish their reputations early by creating and sharing and engaging in ideas in provocative and appropriate ways. These are not easy skills to master. (I'd refer you to Dan Gillmor's new essay "**Principles for a New Media Literacy**" ([tinyurl. com/4b3pos/] for more on that).

My kids need the help of teachers in their classrooms who understand all of this on some personal, practical level. They need teachers who can help them navigate these complex spaces and relationships online that require, at the very least, a different application of traditional skills and literacies. I think as educators we have a duty to do so. You can call it a "fad" if you like, but the reality is that these skills are sorely lacking in our teachers who are suffocating in paper, policies and processes that prevent them from exploring the potential of online networked learning spaces. It's imperative, I think, that we change that. To quote Kansas State professor Michael Wesch, "We [need to] use social media in the classroom not because our students use it, but because we are afraid that social media might be using them—that they are using social media blindly, without recognition of the new challenges and opportunities they might create" (tinyurl .com/9ywq57). To me, that's what 21st Century Skills are all about, teaching our kids to navigate the world as they are experiencing it, not the world we experienced.

Source: tinyurl.com/7dxrbm

PART II

LEARNING IS ANYTIME, ANYWHERE, ANYONE

I Don't Need Your Network (or Your Computer, or Your Tech Plan, or Your . . .)

Right now, our systems and schools still focus primarily on control when it comes to the Internet. If we get to the point where we even allow access in schools, we want kids to use our devices on our networks according to our rules. But what will happen when every student is carrying his own network in his pocket? How do we embrace the ubiquitous access quickly coming into our students' lives instead of trying to resist it? And what do we do for those kids who don't have access?

02 Dec 2009 07:51 am

I've been thinking a lot again about phones and about the disruption they are already creating for most schools (high schools at least) and about the huge brain shift we're going to have to go through collectively to capture the potential for learning in our kids' pockets. A few particular items have kind of come together of late that have been pushing the conversation in my head pretty hard.

First, this kinda cute little **YouTube video titled "Phone Book"** (tinyurl .com/y93vyhm). Not sure who or what it was that led me to it, but it's worth a quick couple of minutes to watch it.

Now take that concept and mix it with these four ideas:

- Apple's next iTouch is coming out with 64GB of memory, and the iPhone won't be too far behind that.
- In the next five years, every phone will be an iPhone. (And let's not forget that there are already over 100,000 apps for that little sucker, many of them with relevance to the classroom.)
- We'll soon be seeing what **Steve Rubel is calling a "dumb shell"** (tinyurl.com/yhqhq8d) that takes the book idea in that video and creates a netbook sized (at least) keyboard and screen that your phone simply plugs into.
- According to NPR, **the Pew Hispanic Center says** (tinyurl.com/ yho7n3h) that there is a definite trend toward phones being chosen over computers as computing devices, especially for those on the wrong end of the current digital divide. (The article makes more sense of that than I just did.)

All of which leads me to ask a whole bunch of questions:

- If at some point in the fairly near future just about every high school kid is going to have a device that connects to the Internet, how much longer can we ask them to stuff it in their lockers at the beginning of the day?

- How are we going to have to rethink the idea that we have to provide our kids a connection? Can we even somewhat get our brains around the idea of letting them use their own?
- At what point do we get out of the business of troubleshooting and fixing technology? Isn't "fixing your own stuff" a 21st Century skill?
- How are we helping our teachers understand the potentials of phones and all of these shifts in general?

And finally, the big kahuna, are we in the process of transforming (not just revising) our curriculum to work in a world that looks (metaphorically, at least) like this: (tinyurl.com/yjs43mf)?

I wonder how many educators look at that picture and think "OMG, puhleeeese let me teach in that classroom!" (I suspect not many.) I wonder how many of them already do teach in classrooms that look like that if we consider the technology in kids' pockets (or lockers) as the access point. (I suspect, more than you think.) The problem is, and I can guarantee you this, 95% of the curriculum currently being delivered in those classrooms would waste 95% of the potential in the room that we could glean from that access.

All too often we get hung up on the technology question, not the curriculum question. Here in New Jersey, every district has to submit a three year "Technology Plan" and as you can guess, most of them are about how many Smart Boards to install or how wireless access will be expanded. Very, very little of it is about how curriculum changes when we have anytime, anywhere learning with anyone in the world. Why aren't we planning for that?

So I'm asking. When do we stop trying to fight the inevitable and start thinking about how to embrace it? Or, as **Doug Johnson so eloquently suggests** (tinyurl.com/yg2krn3), when are we gonna saddle this horse and ride it?

Source: tinyurl.com/ylmv854

What Do We Know About
Our Kids' Futures? Really.

The future is always hard to predict, but from a technology standpoint, some broad brushstrokes are possible. The tools will change, no doubt, but the general affordances of the tools won't as much. Even though this post is more than three years old at this writing, I think the descriptions here still apply. But I wonder how far we've come in helping our students prepare for these eventualities. Can you point to ways in which your classroom or your own practice is moving in the direction of the bullet points below? If not, where is one place to start on that list?

14 Feb 2008 07:33 am

A lot of us (or should I say I?) frame the conversation around Read/Write Web tools in schools in the context of this very blurry future that our kids are entering into, one that despite its lack of clarity is decidedly different from today. In my own case, I tend to frame this through my parenting lens, that it doesn't feel like the system is preparing my kids for their futures very well even though we don't exactly know what that future looks like.

So yesterday here in balmy Toronto, I got asked the question directly: even though we can't be certain about what the future looks like in terms of preparing our kids for it, what, generally speaking, do we know? What general characteristics can we assume in terms of rethinking our curriculum and our practice?

I threw some ideas out, some of which I've tried to articulate below. It's difficult on many levels . . . are we talking about what they need to know in terms of education? Their profession? Environmentally? From a citizenship standpoint? But truth be told, I've been mulling the idea of this post for a while now, so I'd appreciate any sage answers you might be willing to contribute as well. (Come to think of it, this sounds like a potential Tweet . . .)

Our kids' futures will require them to be:

- Networked—They'll need an "outboard brain."
- More collaborative—They are going to need to work closely with people to co-create information.
- More globally aware—Those collaborators may be anywhere in the world.
- Less dependent on paper—Right now, we are still paper training our kids.
- More active—In just about every sense of the word. Physically. Socially. Politically.
- Fluent in creating and consuming hypertext—Basic reading and writing skills will not suffice.
- More connected—To their communities, to their environments, to the world.
- Editors of information—Something we should have been teaching them all along but is even more important now.

There's more, obviously. But I'm curious. What would you add? Or what would you push back against?

Source: tinyurl.com/2pus42

Aggregator as Textbook

Digital technologies allow us to collect and curate a wealth of information about the topics we are interested in online. In essence, we can now build our own texts around any subject, and because of that, we need to become critical editors as we create those flows of information and find value in them. RSS is one of the tools we can use for this purpose. But more important, when we turn from the technology to the practice, how do we do this effectively?

06 Aug 2007 12:34 pm

One of the metaphors I find myself moving more and more to of late is "Aggregator as Textbook." **Google Reader** (tinyurl.com/yk2gcy5) is the place (along with **Twitter** [tinyurl.com/y4lug8] of late) that I head to first every day when I open up my computer, and on an average day, I end up going back there at least 4 or 5 times. It's become an important part of my learning process, because my daily study almost always starts and flows from what's collected there.

That being the case, I've been thinking more and more about my own use of RSS, and trying to reflect on the choices I make in my aggregator. Frankly, I am still amazed that so relatively few people (not just educators) have made RSS a part of their practice, but I wonder if it doesn't have something to do with how disruptive a technology it is when you really think about it. It changes the traditional information structures in fundamental ways, and it forces us to be much more involved with the information we consume. I'm no longer just a reader; I'm an editor who is constantly at work in the process of finding feeds to read, determining what's relevant, trying to connect ideas and patterns, making decisions as to what to do with all of the information I come across.

The technical side to RSS is not that difficult. But I constantly wonder if I'm "doing RSS well" in the way I use it. So, anyway, here are six things I wonder about my own use of what I think is the most powerful of all of these technologies.

What's my optimum number of feeds to read? I've gone between 25 and 250, and now at about 60 I'm still not sure if that's the "right" number. And it's not just a time factor that determines that number, although that has more to do with it than anything else. The scope of topics and a diversity of views also has a lot to do with it.

How do I not become "married" to the feeds I already have? It would be easy to keep the 60 or so feeds that I have for a long time, but I'm not sure that's the best strategy. As new voices appear, as my interests shift, I need to be willing to let some old voices go. That's exceedingly hard, at times, because I don't want to miss anything, and because I feel connected to those teachers on many levels.

Do I rely too much on a handful of feeds? I'll admit, while I struggle reading every feed every day, there are a half dozen or so that I try not to miss. I think of these as the ones that do the best job of culling out the important ideas of the day. In many cases, these people are reading many of the same sources I am. I wonder if this makes it even more difficult to read more widely.

How many individual pieces of information can I realistically make sense of? There are days when I could easily find 50 or so interesting, relevant posts or links to sites, and I wonder if that's always such a good thing. If I were to try to process all of that, will the best filter up?

How do I best organize the information that is most useful? I have a Delicious.com account, and I stow away some snippets of things in various spots. I tag and tag and tag. But this is my most difficult struggle. I've yet to find a really effective way of processing all the ideas and links that make it easy to return to later.

Should I read ideas, or should I read people? Stephen Downes (tinyurl.com/lx35j) advocates for the former, and I can understand why. It's the concept, the exchange of ideas that is important, not the person so much. Still, I find it very difficult to separate the two, and I do think that knowing the person through the writing adds context to the ideas. But, again, reading people also tends to limit the scope and diversity of the ideas, I think.

Without question, my aggregated text requires much more intellectual sweat than the traditional form. And that's actually why I want my own kids to become adept at writing their own texts around the topics they find engaging. I've put together **Pageflakes** (tinyurl.com/8gj39) pages for my kids built on RSS feeds about horses and the Phillies as a way to get them started. But that's just the first step.

So, I wonder, what do you wonder about RSS?

Source: tinyurl.com/2f8265

The Steep "Unlearning Curve"

This has always been one of my favorite posts, primarily because of the title and what it suggests. So much of what we currently do in schools and classrooms is being challenged by these networked learning spaces that we literally have to unlearn much of our traditional practice in order to make sense of it. This process of unlearning is a difficult one because, to some extent, it requires us to step into an unknown that we may not be comfortable with. But it's a question worth asking: What did you unlearn today?

07 Feb 2007 11:17 am

(Cross-posted to "The Pulse") One of the most challenging pieces of figuring out how to move education forward in a systemic way is the "unlearning curve" that we teachers and educators have to go through to even see the possibilities that lay before us. So much of our traditional thinking about personal learning and classroom practice is being challenged by our ability to publish and connect and collaborate primarily because of the opportunities afforded by the Read/Write Web. For instance, in a world where literally any place can be a classroom, we have to unlearn the comforts of four walls that we've become accustomed to. When we can share our work with wide audiences, we need to unlearn the idea that student writing and projects are simply ways to assess what they know.

There is no curriculum for unlearning, and, of course, in many ways it's simply learning to see things differently or to at least be open to it. To me at least, the key is attempting to understand how these technologies can transform our own learning practice (and, I would guess, our unlearning practice as well). If we can get started on that road, it can become much easier to re-envision our classrooms and our schools.

So, with that brief introduction, here are 10 things that I think we need to unlearn:

1. We need to unlearn the idea that we are the sole content experts in the classroom, because we can now connect our kids to people who know far more than we do about the material we're teaching.

2. We need to unlearn the premise that we know more than our kids, because in many cases, they can now be our teachers as well.

3. We need to unlearn the idea that learning itself is an event. In this day and age, it is a continual process.

4. We need to unlearn the strategy that collaborative work inside the classroom is enough and understand that cooperating with students from around the globe can teach relevant and powerful negotiation and team-building skills.

5. We need to unlearn the idea that every student needs to learn the same content when really what they need to learn is how to self-direct their own learning.

6. We need to unlearn the notion that our students don't need to see and understand how we ourselves learn.

7. We need to unlearn our fear of putting ourselves and our students "out there" for we've proven we can do it in safe, relevant and effective ways.

8. We need to unlearn the practice that teaches all students at the same pace. Is it any wonder why so many of our students love to play online games where they move forward at their own pace?

9. We need to unlearn the idea that we can teach our students to be literate in this world by continually blocking and filtering access to the sites and experiences they need our help to navigate.

10. We need to unlearn the premise that real change can happen just by rethinking what happens inside the school walls and understand that education is now a community undertaking on many different levels.

Certainly, there are many others, and I'm sure you have your own unlearning ideas . . . feel free to add.

Source: tinyurl.com/2h62n3

What I Hate About Twitter

Without a doubt, Twitter has become a valuable to ol in my daily learning. It's allowed me to connect with and tap into more information and more people than I could ever have dreamed, probably more than I can adequately handle, in fact. None of these new tools is perfect, and I struggle with each of them in different ways. If you use Twitter, the reactions below will probably resonate. The question, as with all technology tools, is "does Twitter offer enough value to justify the not-so-great stuff?" The answer, still today, is "definitely."

15 Jul 2008 01:50 pm

I've liked Twitter since I first started playing with it last year, but there are some things that are really starting to annoy me about these 140-character "conversations" that we're carrying on there, server issues notwithstanding.

Whether it's some people getting a little snippy from time to time and then other people making a way-too-huge-a-deal about it, or whether it's two very smart people like **Gary** (tinyurl.com/4ckhbhn) and **Sheryl** (bit.ly/oDuGhV) blowing out a Tweet-a-minute micro debate about the state of education in this country, or whether it's people trying to live Tweet hour-long presentations that turn into like 347 updates, I'm finding anything that hints of substance just too scattered, too disjointed to read, even with the wonders of **Tweetdeck** (tinyurl.com/6o482m). It's like trying to eavesdrop on the conversation of a bunch of people with really bad cell phone reception, hearing a part of one response 'til it cuts out into the other. Frustrating.

And I can't help feeling like it's just making all of us, myself included, lazy. We've lamented this before, this "fact" that the whole community is blogging less since Twitter, engaging less deeply, it seems. Reading less. Maybe it's just me (again) or maybe it's my long term attachment to this blogging thing and my not so major attachment to texting, but it feels like the "conversation" is evolving (or would that be devlolving) into pieces instead of wholes, that the connections and the threads are unraveling, almost literally. That while, on some level, the Twitterverse feels even more connected, in reality it's breaking some of the connectedness.

I (we?) blog for many reasons, not the least of which is that I'm sincerely interested in what others are experiencing and I hope to learn from their reactions. When I write here, I can't help but hope that whoever reads it will stop, reflect if they find it relevant, and offer up some wisdom (or whatever else) that will pique my thinking. I hope it becomes a conversation among a group of interested parties that want to test out or build on the ideas. But on Twitter, while I sometimes post silly "I ran five miles" type of check in post for anyone that might be interested, I also find myself writing for just one or two people yet publishing it for everyone to see. And when I read other Tweets directed as a response to another person, it's like I feel compelled to click and dig and sort and try to nail down the context of the "conversation" and then to read it back again to make sense of it.

Look, I love the Tweet links and the "touch 'em alls" and the zen, in-the-moment stuff. But, selfishly, I wonder how much less I might be learning today than B.T. as more of what we care about gets processed in short soundbites.

Not sure why all that tipped for me today, but it just got really painful all of a sudden. Anyone else feeling similar things?

Source: tinyurl.com/5u7x9v

It's the Empowerment, Stupid

As a parent, one of my biggest frustrations is the lack of power schools give my kids to make any real learning decisions (or any decisions, for that matter) on their own. School is in large part scripted; I can pretty much tell you what my kids will be studying two weeks from Tuesday . . . next year. But adult life doesn't look like that. We make decisions about our time, our efforts, and our interests on a regular basis. How can we empower our children to be more self-directed, self-organized learners?

09 May 2007 05:55 pm

Every week, my kids bring home their "Friday Folders" from school, usually packed with paper . . . torn out worksheet pages, handouts from school, permission slips, tests taken, more worksheets, lunch menus, letters from the principal, more worksheets, more tests, an occasional fund raiser, and yet more worksheets. Wendy and I sign our names to much of it, usually in a Monday morning blur, our kids shoving it in front of our faces saying "Just sign it Dad, it's nothing" or something similar when we ask just what it is we're signing. And the next week, that signed paper comes back with another flurry of worksheets and tests and quizzes and god knows what else.

We've been collecting it, all of this Friday Folder paper, growing what's become an enormous pile of it in the corner of our bedroom, a pile that I guess in the eyes of their school in some way represents the learning that my kids have done this year. I'm guessing we're supposed to be proud of all of this accomplishment, this big pile of paper that my kids never, ever revisit as it sits there, growing week by week. Sometimes I look at it and see 1,000 paper airplanes. And sometimes I look at it and wonder if what it really represents is not so much what my kids know as what they have become, a couple of highly dependent learners, enabled by their teachers and their school to produce a constant stream of, of . . . of what? Knowledge? Learning? Busy work?

I was reminded of this by David's post today where he writes about the need for students to become more self-directed, to take charge of more of their own learning in a world where, for the kids who are connected, at least, there is so much more to learn. I know this isn't anything new; we should have been teaching kids that all along. But the fact is that what we've taught them is that the teacher sets the agenda, defines the method, assesses the outcome and controls the whole process. And as David suggests, it's no wonder many teachers and adults in general seem to be waiting for someone, anyone, to teach them instead of taking the initiative to teach themselves; we are most all products of the system.

But I've been giving a great deal of thought to what my own children are going to need to be able to do when they get to where they have to support my wife and I in our old age, and I'm convinced that none of what they are learning now is going to in any way ensure a pleasant retirement for us. They are not being empowered to learn, not being helped to become:

- *Self-learners* who are able to navigate the 10 or 15 or however many job changes people are predicting for them by the time they are 30
- *Self-selectors* who must find and evaluate and finally choose their own teachers and collaborators as they build their own networks of learners
- *Self-editors* who can look at a piece of information and assess it on a variety of levels, not simply believe it because someone else does
- *Self-organizers* who can manage the slew of information coming at them by developing their own structures and strategies for making sense of it all
- *Self-reflectors* who are not solely dependent on external evaluation to drive their decision making and their evolution as learners and people
- *Self-publishers* who understand the power and importance of sharing and connecting information and knowledge and can do it effectively and ethically
- *Self-protectors* who understand where the online dangers lie, can recognize them, and can act appropriately to stay away from harm

Of course, all of this requires a certain willingness to relinquish control, not just of the things we know but of the things we don't know. In fact, that second part is even more important, I think.

The teachers in my kids' school are good people, and I know I'm a tough parent. But the more I look at it, the more I'm convinced that my kids just are not being served by the constant passing of paper back and forth, by a curriculum that's driven by stupid assessments that require answers that may no longer be accurate or relevant by the time my kids need to actually call them up later in life. It's the exact opposite of what they need. And I'm not sure I can sign off on it much longer . . .

Source: tinyurl.com/6m84lb

So What Is the Future of Schools?

Schools as places where we send our kids are not going away any time soon; they are too much a part of our society and culture. But since writing this post back in 2008, I have come around to a different vision of schools as learning centers, where students— especially older students—are able to pursue their own interests and passions with our support. I have no doubt that 20 years from now, many schools will have made the shift. Not all, but many.

05 Dec 2008 08:39 am

So I'm home from Seattle today with some mixed feelings about Microsoft's School of the Future Summit, which was really excellent in some respects but left me wanting in others. The best part without question (and not that surprisingly) were the conversations with folks outside of the session rooms. With 250 or so people from 31 countries, it was probably the most diverse setting (geographically, at least) that I've found myself in. I had some interesting conversations with folks from Norway, Chile, Hong Kong, Sweden, Mexico, Australia, New Zealand, Finland, the UK and others that no doubt gave me a much broader perspective of what the conversation feels like abroad. And it was diverse in ideas as well. My sense is that we're all obviously feeling the pressure to think differently about schools and schooling, but depending on cultures and circumstances, there were a wide variety of approaches to the shift. I heard about models that ranged from kids doing one subject per full day throughout the week (as in math on Monday, science on Tuesday, etc.) to ones that made real use of mobile technologies, to others that were entirely online. (More about that in a second.) I talked to folks who taught in schools where every student had a computer and to others whose few classroom computers ran on dial up, some who were integrating social tools with depth, others who had never really considered them. It was, to put it mildly, a very eclectic and by and large passionate group.

But after a couple of days of listening to speakers like Michael Horn ("Disrupting Class") and Tony Wagner ("The Global Achievement Gap") I can't say that I feel any greater clarity in the conversation around just what schools of the future are supposed to be about. Horn said that in 15 years almost 50% of all courses will be delivered online. Wagner said that we need to reinvent schools but didn't give a very cohesive vision on how to do that. And while it was encouraging to hear Martin Bean of Microsoft talk about teachers and students learning in "media rich, socially connected" spaces as "content creators and knowledge starters," it was less so when he seemed to define the idea of "nurturing powerful communities of learning" simply as creating portals to connect various constituencies. Admittedly, those snippets may not be totally fair to those speakers' larger messages, but they were indicative of the general sense that I got, one that said "yes, we need to do something, but we're not very close to having a cohesive vision around what exactly we should do." Or something like that.

All of which made Rob Paterson's post that came through my Twitter feed in the middle of the conference yesterday so much more thought-provoking. In talking about the pressures facing universities from decreasing budgets and relevance, Rob says It's going to be interesting to see how this unfolds. The web offers a whole new way of restoring this way of learning directly from an expert rather than from an institution.

Rob offers up a vastly different "hang out a shingle" driven model for some seeking learning after high school, one which challenges the diploma driven status quo pretty compellingly. And it really got me wondering (once again) about the relevance of the pretty standard K-12 curriculum and assessments that are driving our systems. As I commented to Rob, I think I'm finally getting to the root of my continued frustration with my kids' education which is the system's inability to help them find and nurture the areas they truly have passion for. It would be nice if the institution were the place that connected my kids to the experts they desired and needed to support their learning, wouldn't it? Again, I know it's more complex than that, but you get the point.

As would be expected, much of the conversation was spent on the barriers to change, and at some point I found myself amazed at how deeply woven the reasons why not are ingrained in our conversations. At one conversation, someone said that many of her teachers didn't feel like they needed to teach with technology at all since their students were doing just fine passing the tests without it. And I wanted to scream (but instead politely said) "then we gotta change the assessments." Nothing in these conversations changed my view that to really change what we do in schools we have to first change our understanding of what it means to teach in this moment. That doesn't mean than we throw out all of the good pedagogy that we've developed over the years and make everything about technology. But it does mean, I think, that technology has to be a part of the way we do our learning business these days. Finally, I think the conversation that most blew me away was the one with Andy Ross, the VP of Florida Virtual High School. They've got almost 1,000 full time staff now and over 20,000 kids on their waiting list to take classes. They can't hire teachers fast enough. Kids can take their entire high school curriculum online without ever meeting a teacher face to face, though there are plenty of phone calls and e-mails. Andy said that their research shows that those kids do better on the standardized assessments than kids in physical schools, primarily because of the deep alignment of the curriculum and the programmed delivery. Now I'm not saying that those are necessarily reasons to move everything online, but it was the one solid vision of a "School of the Future" that I got at the conference. Andy agreed to come on and do a UStream at some point in the near future, and I'll be sure to be posting times and dates in case you'd be interested.

Anyway, just some reflecting on an interesting couple of days . . .

Source: tinyurl.com/6hfjjb

The End of Books? (For Me, at Least?)

I love reading online. I know many of my generation don't, but to me, there's a great value in links and highlights and notes in digital form. And online reading spaces are going to become more ubiquitous, more engaging, more multimedia, and more connected. This story about reading on my iPad using the Kindle app is just one example of this shift, which will explode in the next five years. Is there still value in a book? I hope so, especially this one. (Ironic, I know.) But more and more, eBooks will rule the day.

24 Apr 2010 04:27 pm

So, let me say at the outset that I love books. All my life, I've been a reader of books. I have at least 1,000 of them in my home (on shelves, in stacks on the floor, in boxes in the basement). I have books of every type; novels, non-fiction, story books, picture books and more. Life feels better when I'm surrounded by books.

And I love the fact that my kids love books, that Tucker spent an hour at the public library yesterday, gliding through the stacks, pulling books down, sitting cross legged on the floor, testing them out, that the first thing Tess wanted to do when we moved last fall was organize her books. I totally understand why living in a house full of books is worth upwards of like three grades of ~~literacy in school~~ schooling.

So, with that bit of context, let me try to explain how my book loving brain got really, seriously rocked the other day, rocked to the point where I'm wondering how many more paper books I might accumulate in my life.

Last year, I put the Kindle app on my iPhone and downloaded a couple of books to read. I was surprised in that the experience actually wasn't as bad as I thought it would be. The first book, a great novel by Anita Shreve, was not much different from reading on paper. The story flew by, and other than being surprised when I got to the end (because I didn't know how many pages I had left to go) it was a great reading experience. But non-fiction wasn't so great. If you look at most of the non-fiction books in my library, you'll see they're totally marked up, underlined, annotated and messy. It's the way I attempt to cement in those most important points, and it helps me recall the good stuff in a book more easily. On the Kindle, I could highlight, and take a note, but it just wasn't as useful. The notes were hard to find, and the highlights just weren't feeling as sticky. I wasn't impressed; in fact, it was frustrating. Last week, when I downloaded my first book to my shiny new iPad, things improved. The larger screen made a big difference, creating highlights and typing in reflective notes was a breeze, but I was still feeling the same frustration with the limitations; just because the pages were bigger didn't mean the notes left behind were any easier to find, and stuff just felt too disjointed. I kept searching for a way to copy and paste sections of the book out into Evernote, albeit a clunky process on the iPad, but still worth it if I could make my notes digital (i.e. searchable, remixable, etc.). My searches didn't come up with anything, and I finally turned to Twitter and asked the question there. Ted Bongiovanni (@teddyb109) came to the rescue: @willrich45—re: iPad Kindle cut and paste, sort of. You can highlight, and then grab them from kindle.amazon.com #iPad #kindle. Turns out my iPad Kindle app syncs up all of my highlights and notes to my Amazon account. Who knew? When I finally got to the page Ted pointed me to in my own account, the page that listed every highlight and every note that I had taken on my Kindle version of John Seely Brown's new book Pull, I could only think two words: Game. Changer. All of a sudden, by reading the book electronically as opposed to in print, I now have:

- all of the most relevant, thought-provoking passages from the book listed on one web page, as in my own condensed version of just the best pieces
- all of my notes and reflections attached to those individual notes
- the ability to copy and paste all of those notes and highlights into Evernote which makes them searchable, editable, organizable, connectable and remixable
- the ability to access my book notes and highlights from anywhere I have an Internet connection.

Game. Changer.

I keep thinking, what if I had every note and highlight that I had ever taken in a paper book available to search through, to connect with other similar ideas from other books, to synthesize electronically? It reminds me of the Kevin Kelly quote that I share from time to time in my presentations, the one from the New York Times magazine in 2006 titled "Scan This Book": Turning inked letters into electronic dots that can be read on a screen is simply the first essential step in creating this new library. The real magic will come in the second act, as each word in each book is cross-linked, clustered, cited, extracted, indexed, analyzed, annotated, remixed, reassembled and woven deeper into the culture than ever before. In the new world of books, every bit informs another; every page reads all the other pages. And I also keep thinking about what changes now? How does my note taking in books change? (Do I start using tags and keywords along with adding my reflections?) Now that I can post my notes and highlights publicly, what copyright ramifications are there? How might others find that useful? And the biggest question, do I buy any more paper books? I know others might not find this earth shattering, but this is a pretty heady shift for me right now, one that is definitely disrupting my worldview. And it's, as always, making me think of the implications for my kids. What if they could export out the notes from their own texts, store them, search them, share them? Yikes.

I'm sure I'll be reflecting on it more as it all plays out.

Source: tinyurl.com/3ah953h

No, Actually, You're Out of Balance

This is another one of my all-time favorite posts, primarily because I think it strikes a chord that not many others do. While it's important to find balance in our digital lives, that balance really cuts both ways. I don't think we can be fully effective learners these days if we're not spending a chunk of our time online. If we don't, we're literally ignoring two billion teachers and the sum of the human knowledge we're creating together. In this new learning environment, I think that borders on professional malpractice.

21 Apr 2010 03:50 pm

A quick observation:

Invariably, one of the concerns that educators raise when going down the social technology conversation any length is the "balance" issue, as in we need to maintain a balance between our online and offline lives. The concern is usually raised in the context of too many kids are out of balance, spending too much time on the computer and not enough time engaged in skinning their knees or having face to face interactions with real live humans that will let them practice the important social skills that they are in the process of losing. As a parent, I hear that. Many are usually shocked to find out that I limit the amount of time my kids can just surf around on the web and play games or update their Facebook pages or watch silly YouTube videos. They're 10 and 12, and at that age, and especially now that the weather is warming up, I want them out and about, shooting hoops, jumping on the trampoline, riding their bikes, building forts, helping to mulch the garden (fat chance) and having "fun." That's our parenting choice, and I'm in no way saying it's the only choice or the right choice for every kid or whatever. It's just the way we've decided to approach it. They get their share of time online, and they can negotiate for more if they are doing something creative or productive. But by and large that's what "balance" is for them.

And let me just say that I struggle with the balance thing in my own life as well. I go through phases where I definitely spend too much time on the computer. (Just ask my wife.) I'm currently in one of my stepping back modes, not playing as much on Twitter, trying to spend more time reading and writing deeply instead of in 140 characters (as evidenced by the recent spurt of posts here). Plus I've got basketball practices and games to drive kids to, grass to cut, etc. Sometimes, balance is forced upon you.

But here is the thing: **the reality is that most of those folks who are concerned about kids needing balance are out of balance themselves, just in the opposite way.** They're not online enough, not reading, writing, participating, connecting and creating in these spaces as much as they need to be to fully understand the implications of these technologies for their own learning and for the kids in their classrooms. Lately, when I've been responding to people about the "balance" question, I go with "well, actually, you're out of balance too, you know." I get this kind of stunned silence. What a concept.

I'm all for balance, but if we're going to make that a "concern" around technology use, let's be willing to admit that it goes both ways.

Source: tinyurl.com/23a5nvm

Making Kids "Googleable"

Sharing and being transparent with our thinking and our practice online is a tough shift for many people to make. But, to quote David Wiley, there is no learning, no education, in these networks and communities without sharing. While adults can in many cases see their way clear to putting themselves out there on the web, the discussion becomes even harder when we focus on our students. How important is it for them to learn how to share? What are the benefits, and what are the consequences of not doing so?

09 Apr 2008 07:42 am

I've been doing some informal research of late in my travels, asking some of the principals and administrators that I meet the following question: When you have some applicants lined up for a teaching vacancy, do you "Google" them? Seems a pretty large majority say that yes, they do take some time to see what a standard Google search might pull up about a potential hire. And some even admit to doing a cursory MySpace search to see what comes up. In most cases, they say that the intent is primarily to find out if there is anything negative that surfaces. Almost all of them admit, however, that finding positive things about their applicants, as in portfolios or collaborations or even social sites, does or could make a positive difference in the process.

But then I ask them something along these lines: So if you are Googling people who you might want to teach at your school, what are you doing to insure the kids in your classrooms are "Googled well" when they go for their own interviews? And I don't just mean telling them NOT to post certain things online. I mean what are you doing to help students shape their online portfolios so that when their future employers or future mates run the search, what they find is not just a lack of negatives but a potential plethora of positives? Not surprisingly, the answer is basically "not much."

If we know that it's becoming more and more commonplace to use the web to assess backgrounds and "social capital," and we're doing it in our own hiring processes, when are we going to make that connection in terms of how it relates to our kids' futures?

Would love to hear what your schools do in terms of doing "background" searches on potential teachers.

Source: tinyurl.com/6qjra8

"I Never Knew I Could Have a Network"

A willingness to spend time online and to share our learning empowers us to connect with others from around the globe in ways that didn't exist even ten years ago. These connections are the backbone of the passion-based networks in which we learn today, networks that are literally waiting for our participation. Unfortunately, there are still millions of educators who don't even know these opportunities exist. How can we best ensure that teachers learn about these technologies? How can we model their effective use for others?

25 Feb 2008 11:18 am

That quote from a teacher at one of the schools **Sheryl** (tinyurl.com/36ercf) and I are **working with** (tinyurl.com/c48f92) pretty much sums up the scale of the shift that a lot of educators (and others) are facing these days. And since I heard it last week in one of our sessions, it's stuck with me as a testament to how isolated and how local teaching as a profession still is. At various times, some of us have called these network connections we've created something akin to a virtual staff lounge or pd on demand, and I think most of us ensconced here know the real power is the ability to find others who are equally as passionate about learning and doing in schools and with kids as we are no matter what we teach, no matter what our role. My ongoing awakening to the possibilities of networked learning continues to be one of the most transformative experiences of my life (nothing tops parenting, however) and I simply can't imagine functioning in the world without it.

But I would still venture to guess that 75% (maybe more) of educators in this country still don't know that they can have a network. While most of our kids are hacking away at building their own connections outside of their physical space, most of their teachers still don't have a firm grasp of what any of it means or what the potentials are. And even for many that do know it, there are still legitimate fears and obstacles to creating professional connections online, time and technology at the forefront. If we really come to the point where we want our teachers to learn and teach with technology, we need to do as my **old school** (tinyurl.com/nwvbp8) did and provide them with technology that works, and what **Carolyn's** (tinyurl.com/2ckj48) school has done in terms of beginning to give them the time to learn it and use it well. And, beyond all that, we need **an environment that supports real teaching** (tinyurl.com/43p6nva), not simply curriculum delivery. Unfortunately, very little of that is happening in any systemic ways.

We're in the "Networking as a Second Language" point in teaching, this messy transition phase that is slowly gaining traction where we are beginning to understand what this means but not quite sure yet what to do about it. It's becoming more visible by the day, but it's still hard for most people to wrap their brains around it. It's different; in many ways it flies in the face of what we've come to believe about learning and relationships. The other day, Clarence pointed to

Ulises Mejias' (tinyurl.com/62q9khm) dissertation at Columbia **"Networked Proximity: ICT and the Mediation of Nearness"** (tinyurl.com/yv4h2s) that defines nearness not as something that is dependent on physical proximity but can now be constructed and defined in social, not physical terms. Nearness is inclusion; farness is exclusion. And I like this line especially:

> A more positive interpretation would argue that networked proximity facilitates new kinds of spatially unbound community, and that these emerging forms of sociality are equally or more meaningful than the older ones. Community is thus "liberated," unhinged from space, and can be maintained regardless of distance.

I find that to be true, that in many ways, these connections are more meaningful than the older ones. The passionate learning network of which I am a part is an amazing and important part of my life. The fact that most teachers still have no idea that it is possible is distressing on one hand, motivating on the other.

Source: tinyurl.com/2q2ecf

PART III

THE LEARNER AS NETWORK

The Learner as Network

In many ways, our connections define us as learners, especially today when we can make so many connections online. And we're not just connecting to people; we connect to content and organizations and ideas. No one dictates what connections we should make or what networks we should join. We are driven to those complex choices by our own passion to learn. However we get there, we are active participants in the process, and the process itself is shared to deepen the learning. How are you connecting? How are you adding value in the context of those connections?

30 May 2006 10:17 am

Jeff Jarvis posted one of those **push-my-feeble-brain-to-the-limit posts** (tinyurl .com/bfpo5k) last week which I think has resonance in a lot of ways. It starts with this:

> In the future of media, which is now, everybody is a network. In the past, networks were defined by **control of content or distribution** (tinyurl.com/6nz5tb). But now, you can't own all distribution and content is controlled where it's created.

He writes about how when we work and practice in a transparent, read and write environment, all of us become nodes in much larger networks. (There is a lot of **George Siemens** [tinyurl.com/2uafn] in this.) I love this description:

> Networks are about sharing now; they used to be about control. Networks are two-way; they used to be one-way. Networks are about aggregation more than distribution; they are about finding and being found. Networks are now open while, by their very definition, they used to be closed. You join networks and leave them at will; you can join any number of networks at once and content can be found via any number of networks, there is no practical limit. Networks used to be static. Now networks are fluid.

It's interesting how much this speaks to education, and how far we need to go. We are still about control, not sharing. We are still about distribution, not aggregation. We are still about closed content rather than open. We are static, not fluid. The idea that each of our students can play a relevant, meaningful, important role in the context of these networks is still so foreign to the people who run schools. And yet, more and more, they are creating their own networks, sharing, aggregating, evolving to the disdain of the traditional model of schooling that is becoming more and more irrelevant.

The biggest problem is how few of our educators still cannot relate to this description. They are neither networks unto themselves or nodes of a larger system, and they understand little about what it means to be either in a world that is more globally interconnected. And our students are not only left without models of what it means to be networked, they also get relatively little content that is contextualized through the network. So network literacy, the functions of working in a distributed, collaborative environment (**Jill Walker** [tinyurl .com/3hco284]), is an important aspect of learning and education that precious few of our students get a chance to practice. And it is only by practicing these skills, whether teachers or students, that they can truly be learned.

Source: tinyurl.com/3vyhceo

Social Learning

Learning has always been a social experience, but in online environments, it feels even more so. We can connect with others in real time with just a few mouse clicks, or collaborate on projects at all hours of the night and day, whenever we want to put in the time. But all of it is driven by our desire to learn. That's the biggest reason why these spaces are so compelling, because no matter what we love, no matter what we want to learn, there's someone out there who wants to learn it with us.

25 Jan 2005 11:48 am

My brain hurts. Sometimes there are just too many interesting, intensely profound ideas floating around out there. What did I do BB? (Before blogs . . .)

This is going to be one of those work-it-through, brain dump type posts that probably won't make much sense and rightfully shouldn't even see the light of day without more polish, but, what the heck. We had a snow day yesterday. I'm feeling brave.

My zeal for the potential of weblogs, wikis, RSS etc. is born almost entirely from my reflective self that is constantly amazed at the way these tools have transformed my learning first and my teaching second. This is pure passion for new ideas, for stimulated thought, for dreaming. It is in many ways intoxicating and exhausting. But I really feel like for the first time in my life, I'm getting the most out of my brain.

While I've tapped my intrinsic motivation to learn (which I believe all of us have for subjects that interest us), there is equally intense extrinsic motivation to be a part of a community of learners that is sharing this struggle of ideas with me. They are equally engaged in their passions, and we're able to connect to each other by our blogs and our feeds. This community shares little if any resemblance to the traditional classroom community whose members are motivated neither by the intrinsic joy of pursuing their passions nor the extrinsic pull of being a part of a larger effort to learn.

My learning occurs in the context of a shared construction of meaning. In isolation, meaning-making ends once the meaning is made. In this online community, meaning is never totally made or finished. It evolves and grows, nurtured by the community. Ironically, I used to preach to my Expository Composition students that a piece of writing was never really finished, that they could pick it up again and make new meaning. But we all knew the lie inherent in that promise. It was handed in. It was assessed. It was finished. Not so in this space, however.

I want my own children who are just 5 and 7 to share this passion. But I see it already being bled out of them by well-meaning teachers who are bound by a system that nurtures conformity rather than creativity. I have few choices when it comes to where my kids go to school, but I know I have many choices as a parent in the ways I nurture and support their own learning. My daughter blogs.

She constructs. She publishes. And I want her, eventually, to find her own community. (This goes for my second-child-syndrome son too, by the way.)

Aaron, who has been on a brain bender lately, says "Clearly then, there is an immediate need for educators to find ways to allow students to follow their intrinsic interests in the context of the classroom." This REALLY makes my brain hurt. He's right. But what a huge, huge task. So, my first job is to facilitate that at home, at least, in the bigger classroom that is the Internet. That's where the real learning opportunities seem to be happening.

Source: tinyurl.com/6zdompq

"The Less You Share,
the Less Power You Have"

One of the biggest shifts into networked learning spaces revolves around whom you can trust and how you interact with others. It's not uncommon to hear people say they "Googled" another person to see what the search result would return. And odds are you're being Googled yourself. In many ways, the extent of the information others can find about you determines how trustworthy you are. It's certainly not the only indicator, but more and more, it's an important ingredient. So, how are you becoming more trustworthy, and, by extension, a more powerful learner online?

18 Nov 2008 08:55 am

My friend **Bruce Dixon** (tinyurl.com/qb5y7t) pointed out to me a few weeks ago that if you do a search for "lesson plans" in Google you get almost 9 million hits, which, when you think about it, is a pretty amazing number. Not saying that they are all great plans, mind you, but when you think about the scope and variety of classroom related content that we can mine these days as opposed to just a few years ago.

Yet this concept of sharing content online still seems problematic for a lot of educators. As I travel around talking to teachers, very few of them argue when I suggest that this is still an isolated profession, and I get the strong sense that there is very little articulation around plans, practice or classroom experiences using online tools much less any local digital databases of documents or what have you. When I ask teachers to talk even in general terms about the experiences their students have had previous to arriving in their classes, most sit quietly and scrunch their shoulders. I know, I know . . . there is a time factor involved in doing this, or least a perception of one. But it just seems amazing to me that at this point there is no real shift towards publishing more of what we do, more of what our kids do, not only to expand our own knowledge base but to model for our students the potentials of sharing.

All of this was brought to mind, once again, in an essay by Issac Mao titled **"Sharism: A Mind Revolution"** (tinyurl.com/5la3bp). While I think the ideas may wax a bit too poetic at times, the thesis is powerful: in this world, the less we share, the less power we have. It's an interesting discussion of the challenges to intellectual property and copyright and to the still ingrained perspective that to own and keep private our own best thinking is in some way protective and sustaining of our cultures.

Non-sharing culture misleads us with its absolute separation of Private and Public space. It makes creative action a binary choice between public and private, open and closed. This creates a gap in the spectrum of knowledge. Although this gap has the potential to become a valuable creative space, concerns about privacy make this gap hard to fill. We shouldn't be surprised that, to be safe, most people keep their sharing private and stay "closed." They may fear the Internet creates a potential for abuse that they can't fight alone. However, the paradox is: The less you share, the less power you have.

Mao discusses a lot of the benefits to blogging and sharing, the rewards we can potentially reap, and the positive consequences for the world. And he touches on the implications for education in terms of at least giving our students a leg up in "communication, collaboration and mutual understanding." Not to mention the idea of helping our students to create a digital portfolio that can not only serve to help their teachers get to know them and their passions more effectively but that can connect them to other teachers and mentors who share those passions. And that is power, not only in the knowledge that we gain but in the learning relationships we foster.

Source: tinyurl.com/676gw8

"School as Node"

For more than a century, we've equated getting an education with going to school. What if we began to see schools as just one of many places both online and off where kids became educated? What if schools became the organizing point for connections to many learning opportunities both in our local communities and on the web? To limit our concept of an "education" to something that is delivered to us between September and June, from 8:00–3:00 each day, is not in the best interest of our kids. We need to connect our schools to all of those new possibilities.

27 Sep 2007 06:38 am

I've had **George Siemens's** "Pots, Kettles, and other small appliances of like appearance" post (bit.ly/n8VkwW), open in my tabs for what, three weeks now, and it's been percolating in my brain as I keep mousing across it from time to time, rereading, rethinking. (As a side note, that's an interesting little shift in my practice that the advent of tabbed browsing and sessions management in Firefox has brought, isn't it?) George writes:

> We are at a point of real change in education (k-12, university, even corporate training). We (the edublog community) still carry the baton of change, but if we are unable to conceive a broader vision of systemic change, we'll find ourselves passing the baton to others.

So, that "conceive a broader vision of systemic change" line brought me back (once again) to the shift I think we've been trying to make in this conversation. The one that moves from being about tools and "flatness" to one that begins to really think about and, more importantly, articulate school models and systems in different ways. And even in that discussion, there seems to be two natural camps evolving, those who say reform is next to impossible without totally blowing out the model, and those who feel that we already have some inroads to reform within the current structures, that there are already progressive school models that might begin to point the way. I struggle to find my own way here, for a variety of reasons. I admit that I have little contextual knowledge of this whole debate to bring to the table. My understanding of progressive school reform movements is thin at best, and I'm in catch-up mode. Yet I have two children in a system (not just local) that is badly in need of reform in light of what's coming. Blowing up the model will not work for them (unless we decide to remove them from the system) and, frankly, I don't think there will be a critical mass of folks willing to do this *to the system* for decades to come. Yet I am equally negative on the prospects that schools can meaningfully change in some sort of timely way without starting over. As a good friend of mine who is planning to leave education after 15 years said recently, "I have no hope that the educational system as we know it will appreciably change in my lifetime." He's in his 30s, btw.

Look, I'm a writer. I list to my right. I think in metaphor. So when George says we need a broader vision of systemic change, my mind runs to find words that might begin to piece that vision together in my own brain that might make sense. And as I've been mulling over all of this, of how to best begin to perhaps reframe the way I think and talk about schools that might allow me to think and talk about a "broader vision" of schools, my brain keeps coming back to something that I heard Tom Carroll of **NCTAF** (tinyurl.com/5shx3d4) say last month at that **Institute of the Future seminar I was at** (tinyurl.com/2b699h). And I'm not sure he even remembers that he said it because it was just a few words in a much longer response about the future of teaching, but in the middle of that response he said " . . . school as node . . ."

I wrote that down.

I think for most people, school is still seen as the (THE?) place where kids go to learn. I know that's the way it was for me. Yeah, there was a lot of informal learning that took place on the playground, on Main Street, in the back of cars, etc. But the "real" learning, the important stuff happened at school. It was the center of learning in my life, though I never called it that, per se. But I know that's how my mom saw it. *You went to school to learn because that's where the knowledge was.* And if the teachers at the school were good, they helped you understand why that knowledge was important. And that "vision" worked pretty well for a lot of years. It was pretty easy and consistent.

Problem now is, it's not working any longer. School isn't the only place where the knowledge is. Knowledge is everywhere. You don't have to go to school to get it. And now, because knowledge isn't stuck to a time and a place any longer, knowledge is contextual. It's not one size fits all. The whole idea that 30 kids in a classroom need to learn the same stuff at the same pace at the same time just makes no sense any longer. In this environment, we can't keep thinking of schools as the center of knowledge and learning. Instead, we have to start thinking of schools as a part of a much richer tapestry of an individual's learning and education.

As a node.

Thinking seriously about schools as nodes in larger more expansive networks of personal learning changes the concept of what schools are for. It doesn't diminish their role, but it does reframe it, and I think it places the emphasis where it more appropriately belongs these days: helping students create, edit, and participate in their own networks of learning. (What a concept.) What if we started seeing schools as the places where our students learn how to learn, where, when they are younger, the school may be at the center, but when they leave us, they have built a vast, effective network of learning of their own in which school and schooling is simply one node? Where we've helped them learn how to nurture and sustain those networks to serve them over the long term? Where we've shown them how to leverage those connections in safe, ethical and effective ways? Our roles as educators and systems would no doubt shift away from content delivery toward modeling and supporting each learner's unique journey. And it would challenge us to rethink the ways in which we assess what our students have learned. But that would be crucial and important work, work that some semblance of traditional school structures might actually do pretty well.

But, as **Hugh's great, great drawing** (tinyurl.com/5skf2lb) suggests, we'd have a lot of getting over ourselves to do for that to happen.

So anyway, just some thin early Thursday morning thinking thrown out for comment, push back, hole-poking, name-calling, whatever from a node in the network . . . There is much, much more to consider here, but it is a reframing and some language that at this moment makes some sense to me at least.

(Just as an aside, after thinking about this for a while, I started imagining how school would look as just "a node" in my learning practice right now. As in following "school" on Twitter, or reading the "school" feed in my aggregator, or adding "school" as a friend on Facebook. All of those seem pretty bizarre at first blush, which either means this whole line of thinking is equally bizarre or it speaks to how inelegantly school currently fits into the personal learning network that I'm already a part of.)

Source: tinyurl.com/23epky

PART IV

LEARNING AND LEADERSHIP

Don't, Don't, Don't vs. Do, Do, Do

One of the best ways we can help our students understand the complexities of the moment is to teach them how to be productive, literate learners in networked spaces. All too often, however, schools put more effort into preventing these learning interactions than supporting them. What if we made school more like real life and encouraged our students to take advantage of the opportunities we now have instead of pretending they don't exist? Can we reframe our expectations to tear down the virtual walls rather than building them higher?

20 Sep 2009 08:11 am

Recently, I presented at a school on an opening day for teachers where the first thing that greeted everyone on the table in the lobby was an 8-page Acceptable Use Policy which staff members were picking up as they filed into the school. I picked one up too, and when I had a moment I started paging through it, looking at all the ways in which students (and teachers) could get themselves in trouble on the school network. The middle three pages were filled with an A-Y double spaced list (guess they were saving room for one more rule next year) which spelled out the many transgressions that were not going to be tolerated, things like people shouldn't be harassing one another, going around the filter, accessing shopping sites, accessing any sites that were "social in nature" and, the big one, downloading software to school computers for personal use. And much, much more.

Frankly, I couldn't help thinking that if I was a student in this district, I think I would actually beg NOT to get a computer. Between the filters and the restrictions, I had a hard time imagining what I would be able to use them for in ways that would actually stimulate my learning. I'd rather take my chances with my phone and my computer at home. (About 90% of students in this district had access from home.)

But the other part that struck me was what this policy said about the curriculum in that district. I wondered aloud to some administrators and teachers later if the stiff policies spoke volumes about what they weren't teaching in their classrooms K-12 as their students went through the system. I mean wouldn't it seem that if kids were taught throughout the curriculum about the ethical and appropriate use of computers and the Internet that much more of that policy could be spent going over what students could actually do with the computer rather than the "don't dos" that were listed? At that point, we'd probably have to change the name to an "Admirable Use Policy" or something, but imagine if students walked in on the first day of class, picked up that policy and read things like:

"Do use our network to connect to other students and adults who share your passions with whom you can learn."

"Do use our network to help your teachers find experts and other teachers from around the world."

74   **Learning on the Blog**

"Do use our network to publish your best work in text and multimedia for a global audience."

"Do use our network to explore your own creativity and passions, to ask questions and seek answers from other teachers online."

"Do use our network to download resources that you can use to remix and republish your own learning online."

"Do use our network to collaborate with others to change the world in meaningful, positive ways."

Etc. (Add your own below.)

Now, obviously, that would mean that the curriculum would be preparing students to do that all along. But I'm thinking that if I was a student and I read those "dos" on the first day of school, I'd be itching to get to class.

Source: <u>tinyurl.com/ndvst3</u>

Transparency = Leadership

Leading is not just for administrators; teacher-leaders can play a key role in moving ideas and practices forward. But regardless of one's title, the opportunity to lead can extend far beyond the classroom walls. As I note below, how you do something is now more important than what you do to the extent that if it's not shared, people are less apt to trust your motives. We see this happening all around us, in media, politics, and business. If we want to be trusted and develop community, we must be willing to do our work in the public sphere.

06 Apr 2009 09:16 am

So here is the money question: What two things (and only two) would you tell educational leaders are the most important steps they can take to lead change today? I got that one from a professor at Oakland University last week, and after pausing for what seemed like an excruciatingly long time, I answered "build a learning network online, and make your learning as transparent as possible for those around you." And while I really think the first part of that answer would make sense to most leaders out there, I think the second would have them running for the hills.

It's pretty obvious to me that my own kids are going to be living much more transparent lives than most of their teachers would be comfortable with. I've written and spoken ad nauseum of the need for them to be "Googled well," and I've been thinking a lot lately about a parent's responsibility to start that process for them. (That's a post for another day.) I really do believe that in this moment, however, that schools also have a responsibility to help kids lead transparent lives online in ways that prepare them for the highly complex relationships they will be having in these virtual spaces as adults. But to do that, schools have to get more transparent themselves.

I pulled Dov Seidman's book **"How"** (tinyurl.com/6gk7sfa) off the shelves last week as it speaks so eloquently to this point. I blogged about it **almost two years ago** (tinyurl.com/5r5b62) when it came out, but in light of how things have moved forward since then, it's even more relevant today. While most people see it as a business book, I look at it as a parenting book, one that challenges me to think about how to best prepare my kids for the "hypertransparent and hyperconnected world" in which they are going to work and play. His point is that in that environment, "how" you do something is more important even than "what" you do. If you're not doing it skillfully, ethically, and transparently, you'll be ceding success to those that do.

A big part of my decision making process in terms of who to believe and who to trust stems from how willing a person is to share her ideas, what level of participation she engages in, how ethical or supportive those interactions are, and how relevant she is to my own learning needs. As I said to the many professors in that presentation last week, there is certainly much I could learn from them if they were sharing. But most of them are not.

In this same vein, I have more and more of an expectation of the teachers and especially the administrators in our schools to lead transparent lives. The fact that they are veritably "un-googleable" in terms of finding anything they have created and shared and perhaps collaborated with others on troubles me on a number of levels. First, I can't see for myself whether or not they are learners. And, almost more importantly, I get no sense as to whether or not they are leaders of learners. Whether they are in the classroom or in the front office, I want (demand?) the adults in my schools to be *effective models for living in a transparent world*. I want my kids to see them navigating these spaces effectively, sharing what they know, teaching others outside of their physical space, and contributing to the conversation.

In Gary Hamel's recent piece in the Wall Street Journal, **The Facebook Generation vs. The Fortune 500** (tinyurl.com/c4vld5), he writes

> **Contribution counts for more than credentials.** When you post a video to YouTube, no one asks you if you went to film school. When you write a blog, no one cares whether you have a journalism degree. Position, title, and academic degrees—none of the usual status differentiators carry much weight online. On the web, what counts is not your resume, but what you can contribute.

I totally agree. My kids need to be surrounded by contributors, people who understand the nuances of these spaces and relationships that we interact with on a daily basis. And not only do they need to see contribution, they need to see it done well, ethically, honestly, meaningfully. In other words, this is more than a twice daily update on Facebook or Twitter.

Bringing all of this together, I just started reading the updated version of Howard Gardner's **"Five Minds for the Future"** (tinyurl.com/6dduffa) and there are all sorts of connections to this conversation. Transparency can support all of the ways in which my kids must be able to acquire expertise, act ethically, display creativity, respect diversity, and synthesize and make sense of information. I look at the way my own experiences over the last eight years have pushed me in all of those directions, primarily because I built a network around my passion and I shared most everything I did. I hope I'm being a good role model for my kids in that respect at least.

For most principals or superintendents, however, the idea of making their learning lives transparent is not one that sits too comfortably. It's another one of those huge shifts that is, I think, inevitable but is going to be agonizingly slow in the making. As Seidman asks

> The question before us as we consider what we need to thrive in the internetworked world is: How do we conquer our fear of exposure and turn these new realities into new abilities and behaviors? How can we become proactive about transparency?

Proactive instead of reactive, which is what we're all about when it comes to transparency in schools right now. What a concept.

Source: tinyurl.com/dcss7p

Yeah, You've Got Problems. So Solve Them.

I have no doubt that change is inevitable, but I'm equally certain that our own actions can determine how fast that change occurs. In speaking to schools, I often hear of all the reasons why we can't implement this technology or that pedagogy, usually based on a belief that the problem is intractable or that someone else holds the key to unlocking it. However, I know of some teacher somewhere who has overcome nearly every one of these obstacles. None of them is insurmountable. The question is how willing we are to work to solve them with or without the help of others.

23 Jun 2010 05:00 am

Recently during a presentation a teacher raised his hand and asked what is a fairly common question.

"Look, I agree with most of what you're saying, but I've got kids in my class who don't have the devices, who don't have the access," he said. "What are we supposed to do when every student can't do this?"

I could hear in the voice of the questioner that this lack of access was offered not as a problem to solve but as a reason for inaction, an excuse to maintain the status quo. Normally, the answer I give to that question includes the words "moral imperative" and "digital divide" or some other fairly typical phraseology that tries to honor the challenge, but this time, for some reason, I just looked at the person and said "Great question. How you going to fix that?"

Silence.

I think that's going to be my new strategy, actually, for all of the "yeah buts."

"My students' parents don't approve of these technologies." I hear ya'. How you gonna fix that?

"I don't have time to do all of this." That is a problem. What are you going to do about that?

"My superintendent/principal/supervisor doesn't have a vision for these types of changes." Yeah, that stinks. So, how you gonna help her with that?

We say we want our kids to be problem solvers, but all too often, when faced with the challenges of a changing educational landscape, we don't offer solutions. Instead, we offer excuses as to why we shouldn't solve the problem, why it's better to just keep on keepin' on. And solving these problems is getting easier and easier, actually, as more and more schools have already done the heavy lifting to find and implement solutions. It's not like anyone needs to reinvent the wheel any more. And it's also not like you need a solution overnight, either. Frame the problem, create a timeline and a process, and have at it. If you had say, two years, is there really NO way to solve that access problem?

I know at some level you have to see all of this as a "problem" to solve. You have to REALLY want those kids to have access. You have to look at the world and the ways in which information and communication are changing, and the ways that online communities and networks are becoming powerful learning opportunities, and the move to digital texts and products and look at your school and classroom and have that "Houston, we have a problem moment." But once you do that, **it becomes your problem to solve**, not someone else's.

So yeah, you've got challenges. What are you gonna do about it?

Source: tinyurl.com/268qw9v

"Willing to Be Disturbed"

For real change to happen in our schools and our own practice, we have to be willing to endure some discomfort. Change wouldn't be change without it. Like everyone else, educators and school administrators have difficulty accepting that disruptive feeling. If we're to navigate the coming decades successfully, however, we're going to have to be willing to embrace change, not avoid it. And we'll have to be at ease with constantly revising what we do and who we are in the classroom. That starts with a mindset for growth. How do we empower ourselves to approach the change process differently for the sake of our students and our schools?

14 Aug 2009 12:38 pm

Earlier this week, **I wrote a post bemoaning the ways in which the system treats teachers when it comes to technology** (tinyurl.com/5wonfu5) and I hinted at a different reality for one school I've been working with. Well, that school happens to be **my old school** (tinyurl.com/nwvbp8), the place where I worked as a teacher and an administrator for 21 years before setting out for my current very different existence. And now, due to a somewhat sudden, imminent move to a new house, the place where in all likelihood my own kids will go to high school.

While I love what **Chris Lehmann** (tinyurl.com/62az3t) is doing at Science Leadership Academy in Philadelphia, the problem with the SLA story has always been that it's hard to replicate. Chris is a visionary who was given the chance to build a school pretty much from the ground up, and I think just about everyone would agree that he has done an absolutely amazing job of it. If I could take SLA and clone it, I would. But that's not possible. So, the tougher question has always been how do schools that have been around for 50 or 100 years begin to undertake the real shifts and real changes that are required if they are to move systemically to a point where inquiry-based, student-centered, socially and globally networked learning becomes just the way they do their business? In all honesty, I haven't seen many schools that have *fundamentally set out to redefine* what they do in the classroom in light of the affordances and opportunities that social technologies create for learning. (If you know of any who have a plan to *fundamentally redefine* what they do, please let me know.) There is a great deal of "tinkering on the edges" when it comes to technology, districts that hope that if they incrementally add enough technology into the mix that somehow that equals change. I can't tell you how many schools I've seen that have a whiteboard in every room yet have absolutely nothing different happening from a curriculum perspective. Old wine, new bottles.

That fundamental redefinition is hard. It takes an awareness on the part of leaders that the world is indeed changing and that current assessment regimes and requirements are becoming less and less relevant to the learning goals of the organization. It takes a vision to imagine what the change might look like, not to paint it with hard lines but to at least have the basic brushstrokes down. It takes a culture that celebrates learning not just among students but among teachers and front office personnel and administrators alike, what **Phillip Schlechty** (tinyurl.com/3cy9hza) calls a "learning organization." It takes leadership that while admitting its own discomfort and uncertainty with these shifts is prescient and humble enough to know that the only way to deal with those uncertainties is to meet them full on and to support the messiness that will no doubt occur as the organization works through them. It takes time, years of time, maybe decades to effect these types of changes. It takes money and infrastructure. And I think, most importantly, it takes a plan that's developed collaboratively with every constituency at the table, one that is constantly worked and reworked and adjusted in the process, but one that makes that long-term investment time well spent instead of time spinning wheels. And it takes more, even, than that.

I'm seeing a lot of that happening at Hunterdon Central, my old school. And you can take this perspective for what it's worth since I feel like I played some small part in this process five years ago when we formulated a long-ish term plan for technology that started with piloting a teacher/classroom model for technology when I was there to today, when they are piloting a student 1-1 model (netbooks) for technology this fall. My good friend and former co-conspirator Rob Mancabelli is guiding the work, and he's had amazing success in bringing teachers, supervisors, upper administration, community, students and others into a really "big" conversation about what teaching and learning looks like today, how global and collaborative and transparent it is, and what the implications are for the curriculum and pedagogy in classrooms. This is not tinkering on the edges; this, instead, is a deeply collaborative and reflective process for a small cohort of 30 or so teachers whose kids this fall will all have technology and a ubiquitous connection in hand, a process that encourages them to be creative, to take risks, to make mistakes, and to pursue their own personal learning as well. All of it as a first building block for the systemic, culture change that is hopefully to come in the next few years.

Tuesday, I had the chance to spend a few hours with a part of this group, and I came away just totally energized by the experience. The main reason? Lisa Brady, the superintendent. The cohort group had been meeting throughout the summer, focusing on learning about social networks, on making connections, reading blogs, trying Twitter and Facebook, and thinking about social tools in the context of their curriculum. The teachers come from every discipline, from math to special education to media specialists. And on Tuesday, now as the school year begins to loom large, Rob asked Lisa to address the group and make sure they understood their efforts would be supported. Lisa started by asking everyone to read Margaret Wheatley's **"Willing to be Disturbed"** (tinyurl. com/42s3qhm). I'd urge you to read the whole thing, but the first graph gives you the gist:

> As we work together to restore hope to the future, we need to include a new and strange ally—our willingness to be disturbed. Our willingness to have our beliefs and ideas challenged by what others think. No one person or perspective can give us the answers we need to the problems of today. Paradoxically, we can only find those answers by admitting we don't know. We have to be willing to let go of our certainty and expect ourselves to be confused for a time.

I hadn't expected to try to capture any of what Lisa said next, but as she talked to the teachers, I started writing some of it down. And I started imaging what it would be like if every superintendent walked into a meeting of teachers who are engaged in reaching beyond their comfort zones and learning something new and said things like:

> My question to you is how willing are you to be disturbed? . . . We have to be willing to examine our practice, to be disturbed about what we think we know about teaching and learning . . . We don't really know what we're doing; we're teachers, we're supposed to know, but we don't know everything . . . I'm as unsure about all of this as you are unsure, but I believe we are doing the right thing. It is of critical importance to this organization, of critical importance to our kids . . . Your classrooms are learning labs; we want you to be exploring, looking, analyzing . . . You are fully supported in this work; don't be afraid of what you are doing . . . at this school, we don't change easily, but we change well.

It was really powerful stuff, the superintendent of schools encouraging teachers to take risks, to think differently, to be okay with not knowing, and to know that it's a process, that it's not going to happen overnight. And this is the same type of message Lisa plans to deliver to the full faculty on the first day of school. (The Wheatley piece is being sent to all staff this week.) Already, Central has decided to end the practice of monthly full faculty meetings this year and instead engage in professional conversations around the question "What does teaching and learning look like in the 21st Century?" Since May, all of the supervisors have voluntarily been meeting on a regular basis to study and discuss the shifts around an inquiry/problem based curriculum delivered in networked learning environments. And the teachers in the cohort are archiving and communicating on a Ning site specifically for the work.

Now I know there are some caveats here and not all of this is replicable either. For the last two years, 99% of teachers at Central (3,200 students 9–12, btw) have had their own Tablet PC (for personal and professional use) with wireless connection to an LCD and wireless Internet in every classroom, part of the teacher model that Rob and I started before I left. I would defy anyone to show me a school that has a better customer service oriented technology support plan for teachers and classrooms to make sure everything works. The school has made a fairly substantial financial commitment to the work (with the support of the community . . . budgets pass). And, 99% of kids in the district have Internet access at home.

But despite all of that, what interests me more is the stuff that they're doing that just about any school could do right now: have the conversations, begin to build a culture around change, encourage learning on the part of every segment in the school, and create a long term vision and plan that attempts at least to account for whatever deficiencies or roadblocks currently exist. I see so many schools (SO many) where huge sums of money are spent on technology without any thought of professional learning or thinking about what changes. It's all haphazard, unplanned, unsupported. I talk to so many teachers who just roll their eyes at the newest initiative because a) they haven't had a voice in the process and b) they know the next initiative is right around the corner. There's no thread that binds all of it together, that congeals into a *fundamentally different* vision of teaching and learning. As Chris often says (channeling Roger Schank), "Technology is not additive; it's transformative." But that transformation doesn't come on its own. It comes only when the ground for transformation has been well plowed. Whether we have the budgets or the technology in hand right now, there is little externally, at least, that's preventing these conversations to start, assuming we have real leaders who are willing to be disturbed at the helm.

I'm hoping to follow this story pretty closely this year, but I'm sure it's not the only one. Would love to hear your take on what Central is doing and on other attempts at moving old schools systemically into new places of learning.

Source: tinyurl.com/ksrlzp

"Tinkering Toward Utopia"

We've been introducing piecemeal changes into education for decades now, all of which are meant to transform some part of what we do but most of which are subsumed into the system at the end of the day. These reforms are not what schools require right now. In his great book Leading for Learning, *Phillip Schlechty makes a compelling case for a transformation—rather than a series of small changes—of the school from a place of instruction and teaching to a place of learning. And that, to me, is the key question of the moment. How is learning—the experiential, creative, collaborative learning that we do in our real lives—showing up in schools?*

20 Jul 2009 03:50 pm

During Boot Camp last week, **Sheryl** (tinyurl.com/36ercf) turned me on to Phillip Schlechty's newish book **"Leading for Learning: How to Transform Schools into Learning Organizations"** (tinyurl.com/nohka7) and I had a chance to get through a chunk of it on the cramped, smelly plane(s) to Melbourne. In it, he makes a pretty compelling case that "reform" is really not going to cut it in the face of the disruptions social web technologies are creating and that we really do have to think more about "transform" when it comes to talking about schools. There are echoes of Sir Ken Robinson here, and I've still got Scott McLeod's NECC presentation riff on Christensen's "Disrupting Class" on my brain as well, especially the "the disruption isn't online learning; it's personalized learning" quote. And while there are others who I could cite here who are trumpeting the idea that this isn't business as usual, I think Schlechty does as good a job as I've seen of breaking down why schools in their current form as "bureaucratic" structures will end up on the "ash heap of history" if we don't get our brains around what's happening. In a sentence:

> Schools must be transformed from platforms for instruction to platforms for learning, from bureaucracies bent on control to learning organizations aimed at encouraging disciplined inquiry and creativity.

To that end, Schlechty refers to past efforts at reform as "tinkering toward utopia" and says that if we continue to introduce change at the edges, we'll continue to spin our wheels. He says that schools are made up primarily of two types of systems, operating systems and social systems, and makes the point that up to now, most efforts to improve schools have centered on changing the former, not the latter. Here's a key snip in that case:

As long as any innovations that are introduced can be absorbed by the existing operating systems without violating the limits of the social systems in which they are embedded, change in schools is more a matter of good management than one of leadership. Such changes can, in fact, be introduced through programs and projects and managed quite well by technically competent people who are familiar with the new routines required by the innovations and skilled in communicating to others what they know.

In these cases, while it is sometimes difficult to break old habits, usually after a brief period of resistance, old certainties are abandoned and new certainties are embraced. For example, teachers now routinely use PowerPoint slide shows where once they used overhead projectors and slate boards. The reason this transition was relatively easy to accomplish is that it did not change the role of the teacher. Indeed, PowerPoint makes it easier for teachers to do what they have always done, just as a DVD player is easier to use than a 16 millimeter projector. Moreover, the technical skills required to use a PowerPoint slide show are easily learned and communicated, making the process of diffusion relatively simple.

But when innovations threaten the nature and sources of knowledge to be used or the way power and authority are currently used and distributed—in other words, when they require changes in social systems as well as operating systems—innovation becomes more difficult. This is so because such changes are disruptive in inflexible social systems.

So, from the social media standpoint, the message here is clear. This isn't about doing what you've always done as a teacher or as a school. It challenges those social constructs in the classroom and in the system, and therefore, these shifts are going to be much harder to embrace. Channeling Christensen, he says that existing organizations seldom successfully adopt truly disruptive innovations, and that it's easier to build something new than to change the old. And if you listened to Scott's presentation, you get the idea that the time is ripe for those innovative systems to form and flourish in education. (My question is whether commercial interests will be at the heart of those efforts.)

What I really like about this argument so far, however, is that while the thinking is rooted in the affordances of the technologies, Schlechty also makes the case in the context of citizenship in a democracy as well as a moral imperative that we create citizens who "have discovered how to learn independent of teachers and schools."

Many Americans fear that an inadequate system of education will compromise America's ability to compete in a global economy [hearing Friedman here]. In fact, they have more to fear from the possibility that young people who graduate will lack the skills and understandings needed to function well as citizens in a democracy. Americans have more to fear from the prospect that the IT revolution will so overwhelm citizens with competing facts and opinions *that they will give up their freedom in order to gain some degree of certainty* than they have to fear from economic competition around the world. Leaders should be far more concerned that Americans will cease to know enough to preserve freedom and value liberty, equity, and excellence than they are with how well American students compare on international tests. As numerous scholars have shown, authoritarian leaders and charlatans thrive in a world where ordinary citizens are overwhelmed with facts and competing opinions and lack the ideas and tools to discipline their thinking without appealing to some authority figure for direction and support. [Emphasis mine.]

That resonates with me on so many different levels, on trying to navigate the arguments about global warming, for instance, or in attempting to explain the nuances of the world to my kids who more and more are coming to me with questions inspired by their interactions with online media. The key to this all, to me at least, and a piece that I don't think Schlechty gets, is that much of that now is dependent on our "network literacy" in terms of building our own personal systems of filters and sources that are balanced and open.

The idea that schools become "learning organizations" is compelling in the way that Schlechty describes the shift.

Schools will be places where intellectual work is designed that cause students to want to be instructed and will become platforms that support students in making wise choices among a wide range of sources of instruction available rather than platforms that control and limit the instruction available to them.

That "vision" started me thinking again about what our expectations are for teacher "learning" and the ways in which we might move toward a culture that celebrates and models and makes transparent learning in every corner. One thing that I constantly hear from Sheryl is the idea that we need to see teachers as leaders and as learners, not just teachers. That's such a huge shift here, one that we talked a lot about and struggled with in Boot Camp. And it all makes me wonder what the next decade or two will bring.

Source: tinyurl.com/mqz6v6

"What Do We Do About That?"

Our students' ability to connect with one another is forcing us to rethink just about every aspect of what happens in classrooms. And while this post may be almost six years old at this writing, the questions the teacher asks are still being asked today. The ethics of using online spaces can't be taught in a vacuum, nor can they be taught in a unit in the second half of seventh grade. We're going to have to grapple with some serious questions about cheating and misuse—and we'll have to do that grappling with our students in on the conversation.

25 Oct 2005 12:41 pm

So after spending a great couple of days exploring Monterey and the Salinas Valley area, yesterday started with a keynote (blogged in amazing detail by the estimable **Jenny Levine** [tinyurl.com/9h34lq]) and ended with a white knuckle landing into a windy, rainy Philadelphia just before midnight. (Good to be home.)

But here is the moment that has my stomach roiling (aside from the nasty "snack" the airline gave out): at the end of my presentation, a woman in the audience related the problem with blogs at her school. "The kids are posting questions and answers to tests in between periods so kids later in the day know what's coming. What do we do about that?" My first response was "sounds pretty inventive to me." And I know that some people took that as being flip. But I was being serious. What a great use of the technology, not from an ethical sense, certainly, but from a collaboration and information sense. This is the new reality of a Read/Write world where knowledge is accessible, number one, and knowledge is shared instead of being kept closeted, number two. These kids are finding ways to share the information they need to be successful at what they are doing. Isn't that something we should cheer? (Am I in trouble yet?)

On the plane home, I kept thinking about that teacher's question, about how absolutely relevant and important it was, and how absolutely abhorrent most educators will find the answer. And I wished I'd asked this question in return: How much of what is on that test could those kids potentially find on the Internet anyway? How many of the answers or ideas are already a part of the "sum of all knowledge" that the web is becoming? And why, if the answers are already out here, are we asking our students to give them back to us on an exam? I can understand why we used do this, back in the days when the answers were difficult to find. But today? Instead, why aren't we asking them to first show us they can find the answers on their own, and, second, show us that they understand what those answers mean in terms of their own experience and in the context of what we are trying to teach? Shouldn't we hear what they are saying, that in a world where the answers to the test are easily accessible that the test becomes irrelevant? (And by the way, I'm not saying that all tests are irrelevant in every instance.)

For a long time now, I've been thinking (agonizing?) about what this new landscape means in terms of plagiarism and cheating and ethical use. And I have arrived at the point where it's just so clear to me that it's not the kids that need to change. It's us. We have to redefine what those things mean, because the old definitions just are not reasonable any longer. And please hear me when I say that I'm not advocating that we accept cheating or copying as the way of the world and not work to prevent it. But I am saying that we need to drastically shift our approach to dealing with it. Blocking blogs or websites or Google is not the answer. Asking kids to take tests to see if they have memorized material that they can now find on the web is not the answer. Making two or three or four versions of the test is not the answer.

The answer, I think, lies in teaching our students how to correctly and ethically borrow the ideas and work of others and in demanding that they not just use them but make those ideas their own. That they take the ideas we have tried to teach them and connect them to and show us that they can teach it to someone else with their own spin on it, their own remix. It's so funny that it's taken me until now to truly start to understand what **Lawrence Lessig** (tinyurl.com/r8d3) has been preaching about remix, over a year since I first heard it. It's how learning happens in our own lives. We take the knowledge we need when we need it, apply it to our own circumstance, and learn from the result. We need to say to kids "here is what is important to know, but to learn from it, you need to take it and make it your own, not just tell it back to me. Find your own meaning, your own relevance. Make connections outside of these four walls, because you can and you should and you will. This is what bloggers do (at least the ones who are blogging). And this remix is neither plagiarism or thin thinking. It's the process of learning in a world where, as Lessig says, everything we do with digital content involves producing a copy. This is a profound change from the closed, paper laden classrooms most of us still live in.

And, I'll continue to incessantly beat the drum for educators becoming effective models for how to use all of this information effectively and ethically. Just as we can't teach kids to read well unless we read well, or to write well unless we write well, we won't be able to teach them how to deal with what's ahead if we don't start figuring it out and doing it well ourselves.

So I'm all worked up, and I'm feeling seriously hesitant about putting all of this out there, because I know this is a very, very disruptive line of thinking. Oy.

But if I don't, what am I going to learn?

Source: tinyurl.com/cez9cb

Who's Asking?

One of the things I love about blogging is that it allows me to push not just my thinking but my writing as well. Every now and then I get the urge to take a risk with style and form. This is one instance in which I think it worked, not just to tell the story but to make a point. We're still at a stage where few people are demanding the types of changes being discussed on my blog and many other places, yet there's less debate that the future for our kids will require different skills and literacies than those schools are currently providing. How do we bridge the gap?

22 Aug 2010 04:15 pm

1. The thunder clap makes all of us stop. It's one of those loud, long, rumbling ones, the kind that rolls around in your belly like when you hit one of those hard, deep potholes in your car. It shakes the window panes in the old house, and in that initial crack, we all duck into ourselves a bit, feeling that split second of doom that big summer storms in the Georgia countryside often cause. My kids are throwing the Frisbee in the downpour, and they freeze for an instant as well. I start to tell them to jump inside, here under the porch and wait it out, but before I get the words out they're leaping the puddles, heading in my direction. Smart kids.

The weird thing is that on every porch that I can see on the block, people are out, watching the rain, listening to the thunder. I don't know if they're passing time or just immersing themselves in the strange beauty of the storm, the sheets of water, the muted light, the heaviness of the air. But we're sharing it, my wet, dripping kids, the dog across the street who's sticking his nose out from under the tar paper roof of his doghouse, and the old black man on the opposite corner, folded into his porch swing, puffing on a pipe. We're all watching, and waiting for the break.

Eventually it comes; the thunder rolls are farther away, the rain abates. We pick up the conversation that the noise silenced, the one about our kids and their schools. Miss Frances isn't listening too hard, I can tell, as she gently glides back and forth on her own porch swing. At 91, her concerns are elsewhere. But her son Mike is deep into the troubles of the school system. "They had to cut 15 days out of the school year 'cause they run outta money," he says as he lifts up the brim of his dirt-stained John Deere hat. "They're gonna keep the kids longer during the days, but they just can't afford to keep everything running on those other days." And before he says it, I know what's coming next. Not that it means less educational opportunity for his grand kids. Not that it's a shame to cut the art and music programs to save the football team. Not that there will be fewer teachers, less technology, less learning going on in school this year. Mike runs his hand through his hair.

"I just don't know what we're gonna do with those kids for those extra three weeks outta school," he says.

2. There's no doubt, I'm not from around these parts. I'm just looking for a pack of gum at 7:45 in the morning in Sidney, Iowa, and as I drive into the center of town, in my white Hyundai rental, I'm not seeing a lot of open stores to choose from. It's one of those old, small Midwestern country towns, one with the "we-really-mean-it" city square built around the county government building smack dab in the center of town. I'm looking for some signs of activity, and as I start to curl around the courthouse I spy it; a line of pick up trucks outside a small gas station-convenience store on the corner. I zip into the parking lot and, not seeing any spaces, park awkwardly in front of the double glass doors. I'm running late; I'll only be a minute.

As I get out of the car, through the windows, I see them, a line of men, most north of 60 I'm guessing, coffee cups in hand. They're regulars, no doubt, and before I even step inside, I feel their gaze. They're all jeans and caps and country, and I'm beige khakis, golf shirt and a pony tail. A couple of them nod kindly as I give my own silent, demure "good morning," and after a couple of heartbeats worth of pause to take me in, they go back to their conversation. "It's the schools that should be doin' that," one is saying, and all of a sudden, I'm tuned in, listening over my shoulder as I reach for a pack of Dentyne Ice from the candy shelf beneath the counter. "They're just not teaching it as much as they should be." I step away from the counter, buy a little time by pretending to look closely at the chocolate bars down below, wonder what the system is so deficient in, wondering, maybe . . .

"These kids just don't know nothin' about managing money," he says, and I hear various sounds of assent from the others.

—

So here's the deal with the change that many of us in this conversation are clamoring for in schools: we're about the only ones talking it. The townsfolk down at the corner store aren't demanding "21st Century Skills," technology in every student's hand, an inquiry based curriculum and globally networked classrooms. By and large the parents and grandparents in our communities aren't asking for it. The national conversation isn't about rethinking what happens in classrooms. No one's creating assessments around any of this. And in fact, outside of the small percentage of people who are participating in these networks and communities online, the vast majority of this country and the world doesn't even know that a revolution is brewing.

And, while it's no shocker to say it, that's what makes it really tough to be a leader in schools right now. Because if you're doing your job, you're thinking about doing things that *no one out there is asking you to do*. Which is, after all, what leadership is all about, isn't it? I love Seth Godin's quote from Tribes: "Leadership is a choice; it's the choice not to do nothing." Especially if basically standing pat will get you by. Given the current expectations for "student achievement" and adequate yearly progress, most school leaders can continue to get away with tinkering on the edges and not do anything to really upset the chalk tray. You want to make it into Newsweek's top high schools list? Just keep pumping those AP courses and prepping those test scores. Constructing "modern knowledge" and sharing it with other global learners online? Not finding the check box for that.

I've said it before, you want to lead right now, as an administrator or as a teacher? You have to do both: you have to do all of those things the parents and the town fathers and Newsweek (well, maybe not Newsweek) want you to do, but you also have to start shifting and seeing what the future holds for the kids in your schools, regardless if anyone else can see it. You have to, as the superintendent at my old school Lisa Brady has begun to do, lead your staff and your school community to the place where they understand the need for change as well, a place that's not just about test scores and AYP, but that's about student learning and literacy in new forms, forms that look much different from our own but that will be crucial to our kids' success. You have to be an advocate, wherever and whenever you can, to convince people that while doing both is hard and takes time and effort, that it's worth it, that it's the right thing to do for the kids in our schools.

Because if you're waiting for the conversation in the coffee shop and the porch swing to act, you're going to be waiting a long time.

Source: tinyurl.com/2v34lx9

PART V

PARENT AS PARTNER

It's the Parents' Fault. Not.

We are responsible for educating not only ourselves and our students about the new demands of learning networks, but also our students' parents. And that's a difficult thing to do on many fronts. Parents don't have a lot of time, but they do have an ingrained sense of what education should look like and a clear definition of "success" for their child. They need more than most to understand the shifts in modes of learning. How can we bring parents into these conversations about change in meaningful ways?

26 Oct 2008 06:27 pm

Recently, during a Q & A after a presentation, I had an interesting exchange with a high school principal that went something like this:

Principal: So I just want to give you my take on this.

Me: Sure

Principal: You bring up those examples of kids on MySpace and make the point that no one is really teaching them how to use those sites well.

Me: Yep

Principal: Well, I'll tell you when they learn about that stuff. When I drag them into my office and read them the riot act about what they've been posting to their Facebook pages and they tell me that they never thought other people would look at their pages. They seem genuinely astonished that I could find them.

Me: And whose fault is that?

Principal: Well, I'd like to blame their parents. (Laughter.)

Me: Well, I think it's your fault. (More laughter.) I mean, maybe not you in particular. But whose job is it to educate kids to use those sites well and appropriately? I doubt that most of their parents really have enough of an understanding of what they're doing to prepare them.

Principal: So how do we do that?

I get into some variation of this discussion on a pretty regular basis, but I'm always amazed at how willing school leaders are to admit this reality and how little they are doing to deal with it. There is a solution to this, one that we all know, but one that for some reason few seem willing to implement other than in the guise of a "parent awareness night" or some type of scary Internet predator presentation by a state policeman. For the life of me, I can't understand what is so hard about opening up the first and second and third grade curriculum and finding ways to integrate these skills and literacies in a systemic way. If you want kids to be educated about these tools and environments, then maybe we should, um, educate them.

If we started talking about this stuff in first grade (in age appropriate ways), AND we involved parents in the process by being transparent about our intentions and our outcomes, I'm pretty sure that we could minimize the number of kids who get pulled into the principal's office when they behave badly on their Facebook pages.

Source: tinyurl.com/6zurq2

Dear Kids,
You Don't Have to Go to College

Of the 3,000+ posts I've written on my blog, this has to be the one I'm asked about more than any other. Let me just say that in the five or so years since I wrote it, I'm even more convinced that my kids will have quite a few educational alternatives to the traditional college path. More and more, I see opportunities not only to experience learning in a different way but to create a portfolio of learning that equals nearly any diploma on the wall. I'm not saying college won't play a part; it will just be one of many avenues my kids can take as self-directed learners.

07 Nov 2006 04:37 pm

Dear Tess and Tucker,

For most of your young lives, you've heard your mom and I occasionally talk about your futures by saying that someday you'll travel off to college and get this thing called a degree that will show everyone that you are an expert in something and that will lead you to getting a good job that will make you happy and make you able to raise a family of your own someday. At least, that's what your mom and I have in our heads when we talk about it. But, and I haven't told your mom this yet, I've changed my mind. I want you to know that you don't have to go to college if you don't want to, and that there are other avenues to achieving that future that may be more instructive, more meaningful, and more relevant than getting a degree.

Let me put it to you this way (and I'll explain this more as you get older). I promise to support you for as long as I can in your quest to learn after high school, whatever that might look like. I'll do everything I can to help you find what your passions are and pursue them in whatever ways you decide will allow you to learn as much as you can about them. I'll help you put together your own plan to achieve expertise in that passion, and that plan may include many different activities and environments that look nothing like (and in all likelihood will cost much less than) a traditional college experience. Some of your plan may include classrooms, some may include training or certification programs. But some may also include learning through online video games, virtual communities, and informal networks that you will build around your interests, all moving you further along toward expertise. (Remind me at some point to tell you what a guy named **George Siemens says about this** [tinyurl.com/3tucxws].)

And throughout this process, I will support you in the creation of your learning portfolio, the artifact which when the time comes, you will share to prospective employers or collaborators to begin your life's work. (In all likelihood, in fact, you will probably find these people as a part of this process.) Instead of the piece of paper on the wall that says you are an expert, you will have an array of products and experiences, reflections and conversations that *show* your expertise, *show* what you know, make it transparent. It will be comprised of a body of work and a network of learners that you will continually turn to over time, that will evolve as you evolve, and will capture your most important learning.

I know, I know. Even now you are thinking, "but Dad, wouldn't just going to college be easier?" It might, yes. And depending on what you end up wanting to do, college might still be the best answer. But it might not. And I want to remind you that in my own experience, all of the "learning" I did in all of the college classrooms I've spent time in does not come close to the learning that I've done on my own for the simple reason that now I am learning with people who are just as (if not more) passionate to "know" as I am. And that is what I want for you, to connect to people and environments where your passions connect, and the expectation is that you learn together, not learn on your own. Where you are free to create your own curriculum, find your own teachers, and create your own assessments as they are relevant. Where you make decisions (and your teachers guide you in those decisions) as to what is relevant to know and what isn't instead of someone deciding that for you. Where at the end of the day, you'll look back and find that the vast majority of your effort has been time well spent, not time wasted.

In many ways, I envy you. I think about all of the time I spent "learning" about things that had absolutely no relevance to my life's work simply because I was required to do so. Knowledge that became old almost as soon as it was uttered from my professor's mouth. I think about how much more I could have gotten from those hundreds and hundreds of hours (and dollars) that now feel frittered away because I had no real choice. I want to make sure you know you have a choice.

So, when the time comes, we'll start talking about what roads you might want to pursue and how you might want to pursue them. Your mom and I have high expectations, and we'll do everything we can to support the decisions you make. But ultimately, my hope is that you will learn this on your own, that you will seize the opportunities that this new world of learning and knowledge offers you, and that you will find it as exciting and provocative a place as I have.

Love always, Dad

Source: tinyurl.com/6ysqw3

"So Why Do You Only Give Your Kids 45 Minutes a Day on the Computer?"

Parenting in the digital age can be complex, especially when trying to decide the best use of technology for one's own children. Even at this writing, my wife and I are still unresolved about online gaming, even though the benefits seem to be coming into focus. As much as I use technology, many are amazed that we limit our own kids' use of computers on a daily basis. Since this post, we've loosened the reins a bit, but we're always trying to figure out where the appropriate balance lies.

09 Feb 2009 10:26 am

I've blogged before about how Wendy and I limit the amount of media time that Tess and Tucker get, that we struggle with knowing how much time is too much or too little to be on the computer, watch television, play the Wii or poke around on the iTouch. Most people think that since both of us spend so much time on the computer that we'd naturally let them play all they want. But we don't. In fact, I get the feeling we're more restrictive than many parents, ironically. (Tess swears there's only one other kid in her grade at school that doesn't have a phone yet.)

When I mentioned in passing our 45-minutes-a-day on the computer policy during a recent presentation, I was seriously amazed at how many people came up afterwards (and even e-mailed me later) and asked about that. There was like a whole 'lotta angst going on in terms of people wondering if their kids were getting too much screen time and how we came to the decision to limit our own kids. I had no answer for the first part, and I felt like I stumbled through the second part because to be honest, it's a really complex equation that is going to be different for every kid, every set of parents. For us, I think it's a combination of having two very energetic kids who love to physically play, a reaction to the struggle for balance in my own life, and an expectation that when we're together as a family, we're together as a family that interacts more often than not without media. Frankly, I don't even like it when Tess plays the apps or listens to her iTouch for long periods in the car. But she (and Tucker) can read as much as they want, and they do. We always bring their books with them and we encourage that at every turn. (For some reason, my kids don't get car sick when reading.) Is there a huge distinction? I don't know. Books give us something to talk about. Mario on the iTouch? Notsomuch. And there are exceptions. Tess happens to really like Google SketchUp, and she can almost always get more time if she's making something or exhibiting some creativity. All I know is that we, and I mean we, tend to push back against technology for our kids as much as we embrace it for ourselves. And that is ironic, I know, but that's what we're comfortable with right now when they are 9 and 11. As they get older, they'll get more time, but I know that we'll monitor what they're doing and have lots of conversations about it. When they get ready to start creating and publishing in earnest, we'll certainly help them if that's what they want to do.

Now does that mean that it isn't perfectly ok for some other parents to make other, perhaps more liberal choices about their own kids' media time? Absolutely not. To each his own, and I'm not suggesting to anyone how they parent their kids. I'm also not holding myself up as the poster child for fantastic parenting. (I could tell you stories.) All I know is that's what we're comfortable with right now, that the real cuts and scrapes they get in their physical worlds are more important than the virtual ones at this point, that we are always struggling with it, and that for today at least, I really like who my kids are shaping up to be. They're creative, social, articulate, thoughtful and fun to be around. Most of the time. And I hope some of that, at least, comes from our parenting around technology and media in their lives.

Source: tinyurl.com/bkxbvb

A Parent 2.0's Back to School Dilemma

Few posts that I've written have caused more uproar than this one from early 2010. The 154 comments it received (plus a slew of Tweets on Twitter) covered just about every angle of the "schools need to change" conversation. For some reason, it touched a nerve—in large part, I think, because parents are genuinely frustrated with their kids' schools and don't know what to do about it. This is a dilemma right now for many of us. We love the idea of school, but it's feeling less and less relevant by the day.

09 Sep 2010 08:08 am

Yesterday, **Alec Couros** (tinyurl.com/2g6y93) went "Back to School" to "Meet the Teacher" of his first grade daughter. Here is what he saw:

Here is what he Tweeted:

 courosa **There's so much wrong here.**
about 10 hours ago via Echofon

It reminded me of the night I met Tucker's first grade teacher, and the first words out of her mouth were something to the effect of "First grade is where we learn the rules."

Ugh.

If I'd had Twitter back then, I'm sure I would have Tweeted something similar to this:

 courosa **3 options? 1) Switch schools. 2) homeschool 3) Work with teacher. Disillusioned as I am, I want #3 b/c it's what I know best. Need to try.**
about 10 hours ago via Echofon

Alec's Tweets registered a slew of responses which, to be honest, I found to be a fascinating read, so fascinating that I decided to capture the bulk of them

here (tinyurl.com/28loxup). (Start at the bottom and read up if you want to get the flow of the conversation.) They really are worth the read as they capture not just the emotion of a whole bunch of teacher parents who are met with the same reality when they go to their "Meet the Teacher" nights but also the complexity of what to do about it. It creates a dilemma; do we corner the teacher and give her a new view of the world, look for another class or school, march down to the principal's office, or lay back, do what we can to help that teacher and fill in the blanks at home. We've tried them all, and none of them seem to work very well.

I want my kids' schools to prepare them for the world that I and many of us see them growing toward. *I want it desperately.* (Emphasis mine.) But it's not happening. For Tucker, it means handing in all of his sixth grade assignments in *cursive* (emphasis not mine), and it means another year of 50 lb backpacks filled with less than real world text books and a slew of worksheets that he'll work through and forget. (Tess starts school on Friday so we'll see what her realities are.)

So, while Alec struggles with his realities, I'm once again struggling with mine. And for what it's worth, here's what we'll do to make the best of it once again this year.

1. We write an e-mail (or a letter) to each teacher introducing our kids and ourselves, letting them know what our hopes are, what we'd love to see our kids doing, and what we'll do to support the classroom. We also introduce ourselves, and talk a little bit about what our worldview of education looks like. Finally, we offer to continue that conversation and help make it a reality in the classroom in whatever way we can. And we cc the principal and headmaster (since Tess is in private school).

2. We co-school as much as we can. I found the Tweet by @dschink to capture it pretty well:

"We've always considered public school ed our kids receive as supplemental to the ed we provide at home so we don't go crazy about it."

Problem is, at least in our case, co-schooling is pretty scattershot, not as deep as I'd like it to be, and frustrating at times for our kids. In other words, I feel like we do our best to engage our kids in the bigger conversations, but it's the reality of both parents being self-employed that it doesn't always work that well.

3. We opt out when we can. I've written notes to teachers in the past when my kids get the first 10 problems of the homework right excusing them from the next 20 same old same old problems on the worksheet. Gets interesting responses sometimes. Also, this year, we're 90% sure we're going to have Tucker opt out of the 6th Grade NJ ASK assessment. Enough is enough.

4. We occasionally send links with resources to specific teachers and cc the principal.

I'm sure we could do more, but my radar to meddling parents may be a bit too sensitive having been in the classroom for 20 years previous. I know how difficult it is. I don't want to make it more difficult, but I do want to try to strike that balance. Hard sometimes.

Wondering what other strategies might be working for you?

Source: tinyurl.com/3ygsesk

A Summer Rant: What's Up With Parents?

Every now and then, I dial up a real diatribe on my blog. This is one of those posts. I want to emphasize again that my frustrations and most of the frustrations of the parents I talk to are not about teachers as much as the systems those teachers teach in. And most don't necessarily want schools to experiment on their kids. But I still wonder why parents aren't demanding more in terms of engagement and learning from their schools. The distinctions seem pretty clear—to me, at least. You?

09 Jul 2010 08:20 am

<rant>

So here's the question I'm grappling with: why aren't parents more angry about the education their kids are getting? I know, I know . . . it's the same system they went through, most schools are getting over the traditional bar, the whole technology is changing learning thing isn't dinner time conversation . . . I get all that. So what?

Humor me. Bring some imaginary sets (or onesies) of parents into a room and ask them these questions. What kind of responses do you think you'd get?

- Do you want your kids to be problem solvers?
- Do you want them to be able to work constructively with others to create useful stuff?
- Do you want the things they create to contribute to the community?
- Do you want your kids to be able to distinguish between relevant, truthful information and the alternative?
- Do you want your kids to be creative, imaginative and curious?
- Do you want your kids to work on their own, to self-direct their own learning?
- Do you want your children to use technology to learn and create?
- Do you want your kids to be passionate about learning?
- Do you want your kids to be engaged in school?
- Do you want your students to learn from/with different cultures?
- Do you want them to be independent?
- (Add your own here.)

I'm thinking few if any parents are going to say, "um, no, I don't really want that for my child." Right? Ok, so now ask them, "How's your kid's school doing with all of that?" Unless I'm just totally being delusional here, I think they'd struggle when pressed to assess the problem solving, collaboration, information sifting skills et al. that their children are getting. I know I do. I mean, where is the grade for all of that? How many parents actively try to make qualitative judgments about all that stuff based on the conversations and work that their kids bring home?

Not many.

The other night, I asked those kinds of questions to some parent friends of ours as we went deep into the night talking about education. These were really smart, caring folks who absolutely wanted all of the above for their own children but didn't really know if it was happening. I got the sense that they had an implicit trust in the system to do right by their kids, and that the grades their kids received pretty much told the story of their education. They struggled with those questions. Not to say they weren't frustrated with some of the things that happened in the school. Not to say they were always happy. But they seemed powerless, even resistant to change it.

They weren't pissed. (I am.)

We read everywhere that US school kids are lagging behind, and we all go through the requisite amount of hand wringing and worry. And I know that in the mostly white and privileged communities in which most of us live (have you really looked at the picture/avatars in your Twitter list lately?), it's easy to say that it's the other kids that are lagging, not ours. And I also know that for many, many people, just being able to go to school and do well on the traditional tests is an amazing blessing. I'm not suggesting this isn't complex.

But if we really believe in the value of all that problem solving, collaboration, self-direction, passion stuff, and we take an honest look at what the current system values by what it assesses, it's hard not to see the gap. I know, we get the assessments we can afford. I know at the end of the day, assessing all of that really important stuff doesn't fit the "easy" model we have for schools right now. But what I don't know is why there isn't more urgency coming from the home. Do parents think that all of that stuff is just folded into the class grade somehow? Really? Or is there a fundamental reality about all of this that I can't see (or maybe I'm not willing to admit?).

What's up with that?

</rant>

Source: tinyurl.com/2fu43ws

Owning the Teaching . . . and the Learning

I love this quote, which describes the options teachers have in the face of change: "complain, quit, or innovate." The question is, "What can we do to make that last option more of a reality for teachers?" Not to say that there aren't a lot of people taking steps to genuinely change their classrooms, but we need more. We need you. At a time of great change in the world outside of schools, we need more people to make change happen inside. What will it take to stoke the flames of innovation in your classroom?

03 Nov 2006 10:16 pm

I've been growing more frustrated lately and I'm feeling more pessimistic about the prospects for any serious change in how we as an education system see teaching and learning, and I think I've figured out why. I hate to generalize, but the thing that seems to be missing from most of my conversations with classroom teachers and administrators is a willingness to even try to re-envision their own learning, not just their students. Many will say that they understand to varying degrees the changes that are occurring, that the web is in many ways rewriting the rules of communication and socialization, that the world our students enter when they leave us will be much different from the ones we ourselves were prepared for. But it feels like there is this unspoken belief among most that we can deal with these changes without changing ourselves. And that is a huge problem.

Lots of teachers I talk to want blogs and podcasts and wikis. Without question, there are thousands of teachers, tens of thousands in fact, who are already using the tools with their students. I see new examples every day. But I'm still bothered by the fact that very, very rarely do I see new pedagogies to go along with them that prepare students for the creation of their own learning networks. That allow them to take some ownership (or at least envision the possibility of it) over their learning. That help them learn self-direction and get them to stop waiting for someone else to initiate the learning. And even rarer is to find one of those teachers exploring his or her own learning through the tools.

More than anything else, I think, teaching is modeling. As a writing teacher, I wrote with my students. As a journalism teacher, I wrote for publication with my students. As a literature teacher, I practiced and modeled reading for my students. Modeling is teaching, and never has that been made more apparent to me than when my own children act out and reflect my own bad behavior back to me. (It happens more than I like to admit.) My own kids, it has become clear, learn less when I talk, more when I do. And so it is with me.

We go back and forth in this community about whether teachers who use blogs should blog, or podcast or read RSS feeds. I've always hesitated to come down on one side or the other in that debate for a variety of reasons. But it's become clear to me that the answer has to be yes. If you are an educator,

I think you have little choice but to choose option 3 in the **Marco Torres** (tinyurl .com/68j25p5) mantra: "You can complain, quit or innovate." I know in many ways it stinks to have to be an educator at a moment in history when things are changing on a glacial scale. But what you signed up for is preparing kids for their futures. You have little choice but to deal.

Why won't our kids be as well served if we don't change ourselves? I mean we're all products of the system, right? We all did ok. Things were changing when we went through school, right? Um, no. Not like this.

Our students will by and large have the ability to learn anything, anywhere, anytime (if they can't already). The level of their collaboration and connections with colleagues and peers in online environments will be of a type that is hard for most of us to imagine (myself included.) The information and knowledge that they will be awash in will require skills and literacies that most of us **simply do not have** (tinyurl.com/4yfcjtm). Their futures (and to some extent their "presents") look very little like our vision of what it means to be educated. (And if you don't believe that, spend some time reading "The Education Map of the Decade.")

And so here is the friction: Recently, I had a teacher tell me that she spent about 10 minutes a day online and that frankly, that was quite enough. She said that she's not going to sacrifice the other things that she already does in her life to spend more time on the Internet. I wanted to say, as **Yochai Benkler says in the Wealth of Networks** (tinyurl.com/3syo2zq), you have the "greatest library in human history" at your fingertips. You have a billion potential teachers. You have an opportunity to learn in ways that you or I could not even have dreamed of when we were in school. And you have an opportunity to shepherd your students into a much more complex, much messier, and much more profound world of learning in ways that will help prepare them more powerfully for the world they face.

Many of our kids are already doing this without us. Many of them have much more of a clue of what it means to learn using these tools than we do. Imagine if we could teach them to leverage their connections even more powerfully, if we could show them how powerful they are in our own learning. That we are not just engaged teachers but engaged learners. That we're not afraid of what's ahead because we know how to learn.

Surely, that's worth more than 10 minutes a day.

But the litany of reasons why this can't happen are on the tips of too many tongues. Today, in our parent conferences, I asked my daughter's teacher if there were opportunities for her class to work on extended projects, projects that in the end would have a purpose beyond the grade and the classroom. Projects that, to quote Marco again, would "have wings." The response I got was this: with all of the objectives that must be met for the state tests coming up in the spring, there just isn't time for it. When I asked my son's teacher whether she had read his blog, her answer was that blogs were blocked at school and so, no, she hadn't.

And so I am frustrated, and I am wondering what will it take to make our classrooms places of learning rather than places of teaching. And I'm wondering if **teaching really is dead** (tinyurl.com/3jxc8sx). And I'm wondering, like the survey question from a few days ago, what classrooms might look like 10 years from now, if they will be fundamentally different from what they are today.

My guess right now is not much.

Source: tinyurl.com/6489u5

The Ultimate Disruption for Schools

As it stands right now, the big picture of the education world is this: the web is disrupting the fundamental function of schools. This doesn't mean that schools will go away or that teachers will be rendered unimportant. But make no mistake, this is a big shift. This particular post captures the extent of both the challenge and the opportunity we have right now. And it's an opportunity that we have to take advantage of, for ourselves and for our students.

11 Dec 2008 01:56 pm

So sue me if sometimes I get too smitten with those who write compellingly and with vision about what all of this connective learning stuff means for the long term, but I love to read stuff that makes my head shift and hurt at the same time. Case in point is this **post by Mark Pesce titled "Fluid Learning"** (tinyurl. com/45x9bpk) which I read first last week and have reread a few time since. I know it's not free of holes, but I have to admit that the picture he paints of higher education in the near future resonates with a lot of my own thinking, and it's got me ruminating even more deeply on what all of this means for my 9 and 11 year old in terms of what their education is preparing them for.

Start with this:

> The computer—or, most specifically, the global Internet connected to it—is ultimately disruptive, not just to the classroom learning experience, but to the entire rationale of the classroom, the school, the institution of learning.

That will at least give you a sense of where he's going with this, and I'll give you the briefest of synopsis with the hope you'll read the whole thing.

He starts with the story of **RateMyProfessors.com** (tinyurl.com/h4pd8) and the influence it's having on decision making by students and universities in terms of the courses they take and the people they hire respectively. During a PLP session last night where we were talking about this, Robin Ellis chimed in that her son had relied heavily on the site throughout his college career, and I'm sure others would attest to that as well. (I pinged a few of my former students on Facebook and they all were avid users.) While this wasn't the original intent of the guys who created the site

> . . . knowledge, once pooled, takes on a life of its own, and finds itself in places where it has uses that its makers never intended.

But what I'm really chewing on is the idea that we can do much of what higher ed offers on our own these days. That, I think, has huge implications for my kids and for the way we prepare students for their learning futures. Pesce asks

> Students can share their ratings online—why wouldn't they also share their educational goals? Once they've pooled their goals, what keeps them from recruiting their own instructor, booking their own classroom, indeed, just doing it all themselves? . . . Why not create a new kind of "Open University," a website that offers nothing but the kinds of scheduling and coordination tools students might need to organize their own courses?

And, to really push that thought:

> In this near future world, students are the administrators.

Whether or not my kids decide to go to college, the question for me right now is shouldn't my school system be preparing them equally as well for a world where traditional college is not the only route to academic success? Shouldn't my kids get some concept of how to gather their own information, find their own teachers, develop their own collaborative classrooms and write their own curricula? I mean at the very least, shouldn't we let kids know that is an option these days?

And as the role of students changes, so too does the role of teachers and classrooms. Teachers are mentors and facilitators (not a new idea, I know) and classrooms can be anywhere.

> The classroom in this fungible future of student administrators and evolved lecturers is any place where learning happens. If it can happen entirely online, that will be the classroom.

Pesce ends with four recommendations. First, "Capture Everything":

> This should now be standard operating procedure for education at all levels, for all subject areas. It simply makes no sense to waste my words—literally, pouring them away—when with very little infrastructure an audio recording can be made, and, with just a bit more infrastructure, a video recording can be made.

Second, "Share Everything":

> The center of this argument is simple, though subtle: the more something is shared, the more valuable it becomes.

Third, "Open Everything," not just using open source, but creating "device interdependence" and in taking down the filters:

> Education happens everywhere, not just with your nose down in a book, or stuck into a computer screen. There are many screens today, and while the laptop screen may be the most familiar to educators, the mobile handset has a screen which is, in many ways, more vital . . . Filtering, while providing a stopgap, only leaves students painfully aware of how disconnected the classroom is from the real world. Filtering makes the classroom less flexible and less responsive. Filtering is lazy.

And fourth, "Only Connect," connecting students to their teachers and their peers:

> Mentorship has exploded out of the classroom and, through connectivity, entered everyday life. Students should also be able to freely connect with educational administration; a fruitful relationship will keep students actively engaged in the mechanics of their education . . . Students can instruct one another, can mentor one another, can teach one another. All of this happens already in every classroom; it's long past time to provide the tools to accelerate this natural and effective form of education.

I know this last is a huge challenge for teachers and schools, but the reality is that we can connect to our teachers any time we like these days, and there are always teachers available. It's just another way in which the traditional classroom is looking less and less like the real world.

Read the whole thing and, if you like, come back here and push the conversation in terms of K-12. I'll write more about this later, but I am approaching the breaking point in terms of what my kids are getting at school. I've got to figure out a better way . . .

Source: tinyurl.com/5vsoz5

PART VI

THE BIGGER SHIFTS: DEAL WITH IT

The Bigger Shifts . . . Deal With It

It's clear that much of the friction schools are experiencing with online tools is cultural and that changing a culture is probably the hardest work of all. For all the potential of the web for learning, there is a corresponding faction of not-so-great people, sites, and content that can wend their way into our online lives. But the solution to the darker side of the web is not to turn the whole thing off or filter it into uselessness. How can we begin to teach kids, and ourselves, to deal with the realities of an online life?

13 Jul 2006 03:17 pm

I just finished my three-day stint at the **High School's New Face** (tinyurl .com/3be23kh) conference and I have to say I'm impressed and encouraged by the conversations here. Last night at dinner, people said their eyes had been opened and that for many at least, they felt had a real chance to make some changes at their schools. There was a lot of excitement about the technology, about the willingness to consider different models of schools (like The Met) and about the strategies for bringing those changes about.

But there was a moment in today's last workshop session that captured the road ahead for this group and for the others that have gone down this path. I was just finishing up an hour on podcasting, showing them how to save **Skype** (tinyurl.com/26m3w34) calls and mix them with music and other mp3 files, and showing them how easy it is to create an audio post and a podcast with **Odeo** (tinyurl.com/3ouc4zb). It was great, I mean, they even broke into semi-spontaneous applause at how easy it all was, and it was obvious they were getting juiced by the potentials. Life was good . . .

. . . until, of course, someone noticed that the number 9 listed podcast on Odeo is called "Open Source Sex."

"So much for that," the teacher who noticed it said. "They'll never let this site through." Talk about air going out of the balloon. I think I rescued it by reminding them how easy it is to do this with Audacity and **Our Media** (tinyurl .com/6ypat), but the point was clear. We may have great ideas and be thinking differently about learning, but it ain't gonna fly when implementation time comes.

And so there it is. Another one of those nasty little truths about all of this. The biggest shift is not the technology, not the practice, not even the implementation. It's the cultural, social shift that moves us from the idea that we must prevent our kids from seeing and engaging with this "stuff" to the idea that says, look . . . it's a different world . . . they're going to find sex and porn and bad stuff and bad people no matter how hard we try to keep them from it, but when we weigh that fact against the incredible learning potential that the web provides, we're going to choose to educate rather try to block and filter it all.

What kills me most about all of this is that I have yet to see anyone cover the eyes of their kids when they go into a magazine store and every skinny, big-breasted super model or super actress is right at eye level, or change the channel when scantily clad women dance provocatively in front of half naked, muscle bound men in the name of **selling beer** (tinyurl.com/69vm9ur) or music or whatever else, or stop them from going to movies filled with violence, abuse, objectification and the rest. Why is there no outrage over that? Is it because that's done within full view of parents? Is it because we've just become so inured to it that we don't see it. (I doubt that.) Is the web different because the kids are at the controls? What is the mentality that says seeing it all around us in public is somehow less "damaging" than seeing a word on a website somewhere?

Just to be clear, I don't like it at all that this is a much more difficult, complex world for all of us to have to navigate. I've said this before, but every time I think what my own kids see and hear just in the course of their normal day, I get just totally disgusted with what we choose to subject them to as a society. But that's the reality. And I deal with it by pointing it out at every turn, by making sure they have the editorial skills they need to deconstruct the image and get to the message and understand the motives behind it. And to frame all of it in a larger context of what beauty and health and happiness really is. I can't keep them from that bad stuff. But I can help them understand it and to at least have a chance of making good decisions about it when they are confronted with it.

But we're just not willing to deal with that in schools, it seems. Why is that?

Source: tinyurl.com/f3yxz

Failing Our Kids

Technology aside, many of our kids face more general inequities when it comes to the schools they attend and the communities they live in. This post is an attempt to capture some of those divides. I'm not sure, to be honest, the extent to which computers and online networks can help our neediest children, though I do think they should be part of the discussion. Access to what's online is a requirement for being educated in this century. We're leaving many of our kids behind.

14 Apr 2009 10:47 am

My nine-year-old Tucker plays AAU basketball for a struggling inner-city team about 30 minutes from where we live. His teammates call him "Shadow" and most times we are the only white family in the gym for games and practice. We (mostly my wife Wendy) haul his (and his sister's) butt down there three times a week for a couple of reasons, first and foremost because we want him to see that a large chunk of the world looks little like the un-diverse, rural space in which he's growing up, and, second, because the basketball is just grittier, tougher, faster, played at a different level than in these parts. The gym in which his team plays is 2/3 the size of regulation court with blue-padded stanchions that jut out from the sidelines and become part of the game, and dim fluorescent lighting that depending on the level of sunlight filtering in from the grimy skylights makes the basket a dark target. It's a no blood-no foul type of game they play, the fundamentals of which are no look passes and under the basket scoop layups which even on a 10-year-old level are both beautiful and at the same time difficult to watch. For most of these kids, basketball is a respite from the difficulties of their lives, lives that are surrounded by poverty, violence and drug use. There are gangs in the middle schools, absent fathers, job layoffs and more, so whenever these kids get the chance, they play, and play, and play some more. And my kids try to keep up.

Tucker has made some fast friends with his teammates. They are sweet, respectful, fun kids to be around. The last couple of weekends, we've hosted sleepovers, or more aptly, shootovers as most of the time the sounds of basketballs being pounded by the hoop at the end of the driveway echo through the house. But we've also been doing some "field trippy" sort of stuff. A couple of weekends ago, Wendy got their parents to give them a day off of school to go to a **statewide GreenFest** (tinyurl.com/26zac8u) to have fun but, as is my wife's way, to get them thinking about the environment. They saw solar cars, learned about organic foods and, at one point, got a lesson on worms. Each of them got a container with some compost, a few poop generating worms, and instructions on how to use them to create great fertilizer for plants. It turned out that for two of the three kids that Wendy spirited off with, it was the first time they had ever held a worm. In the course of the few days they were hanging around with them, we found out all sorts of stuff about their lives and about what they knew about the world, which was, not too surprisingly, not much. At one point when Wendy asked one of them how many people he thought were in the world, he answered "10,000." The next weekend, we went to **"Ringing Rocks"** (tinyurl.com/d3q2qx) which is this strange little geologic enigma near us, followed by some first-time skipping of stones in the Delaware River near our house. It was an interesting few days of learning for all of us.

There is no doubt that these kids face some pretty difficult futures as a result of circumstances not of their making. It's pretty obvious they are behind in terms of what they know about the world and their ability to express it well. That's not an indictment on their schools, per se, as much as it is the inequality that exists in this state and others between the education of the haves and the have nots writ large. But while they say they get "Bs" in school, I can't help but wonder what that means. No doubt, their learning lives are aimed at what's on the state assessment, yet they are behind in reading and writing and math. And to be honest, I'm not sure the system can overcome the difficulties present in these kids' lives from the start. I don't think the answer for them is longer school years or teachers getting "merit pay" (or battle pay) as much as it is a fix for the societal problems that surround them. Yet in this moment of steep budget cuts and layoffs, those fixes don't seem to be on the horizon for them any time soon.

But it's not just them. Last week I was on a panel with the state assistant commissioner of education where she told the story of seeing the "new" digitally published third-grade "U.S. States" projects, the ones we all did as kids, taking a state of the union and pasting the state bird and state flag and state flower on top of a map with some interesting statistics around it. She asked one young man who did New York State to talk about his slide and he read off all of the stuff. When he got to the population part he said "and New York State has over 19 million people," and she responded with "Wow! Is that a lot of people?" He looked at her for a moment and said, "you know, I really don't know." It was a great example of the context and value that information loses when we fail to teach meaning over memorization.

For Tucker's friends, for that kid learning about New York, for a lot of kids in this country, it becomes obvious very quickly that we are failing them. Like I said, I know it's more complex than just blaming the schools and the teachers, which seems to be de rigeur these days, btw. Which is what is so disheartening about the rhetoric that continues to come out of Washington around education; there's nothing really new. Nothing bold. Nothing that makes me feel like we've turned any corner on any of this. We're arguing about the same old ideas and writing about the same old shifts when the reality is that the lives of those kids on Tucker's team haven't changed a bit from all the bloviating going on.

Not suggesting I have the answer here. My frustration just gets more acute when faces and smiles and hook shots come with the statistics.

Source: tinyurl.com/clmeez

Why Blogging Is Hard . . . Still

I've been blogging for 10 years, and while sometimes it feels like the words just flow through my fingers to the keys, other times notsomuch. Any writing that will be shared with a potentially global audience requires some "intellectual sweat" and careful consideration. But having said that, if I don't write, if I don't publish, if I don't share, I don't learn. Aside from my children, my blog has been the most transformative thing in my life. In many ways, I blog, therefore I am.

09 Jan 2009 10:46 am

So at some point in recent weeks the blog-post-o-meter rolled through 3,000, and if I'm even close in my estimation that the average length of posts over the last seven and a half years has been around 3–400 words, that suggests about 1 million words of writing and reflecting and thinking here. That's a pretty staggering number in my feeble brain. You'd think that after all of that output this publishing thing would be almost as easy as breathing.

Well, it's not.

I'm reminded of this because of conversations we've been having of late with **team leaders in PLP** (tinyurl.com/c48f92). While the successes are many and impressive, a good number of people still find the thought of publishing to an audience, even a relatively small, private audience of like-minded souls, to be too daunting. It's just way outside their comfort zone, and they just believe that their contributions would either not be relevant, interesting or useful. It's hard to nurture these folks, to convince them to take small steps, to help them see the potential upside. And I really believe that there is an upside to sharing what you know and do with others; it's the foundation for building learning networks.

But here is the thing: no matter how you slice it, blogging is a risk. And it's a risk not just because you are putting yourself out there for the world, but because unlike many other types of writing that we do, it's unfinished. At least that's the way it feels for me. I don't KNOW very much for certain. But blogging isn't about what I know as much as it's about what I think I know, and I find that to be a crucial distinction. For me, it's the distinction that constantly makes this hard. It's also the distinction, however, that makes blogging worth it. The one thing that a potential global audience does more than anything else is create the opportunity to really learn through writing in various texts, through the conversation and feedback that ensues. I say this all the time, that while a lot of my learning occurs in the composing of the post (or whatever), most of it occurs in the distributed reactions (when they happen) after I publish.

One thing I do know is that when I write with a humility of not knowing I get a lot more learning in return. That plays out in my reading as well. I am not the greatest commenter on other people's blogs (though I am working on that). But I find I am much more compelled to comment on posts where the author is obviously testing unfinished ideas. Where that person is not simply saying "this is the way the world is." I find those types of posts less compelling.

And, obviously, the other risk is that my "thin thinking" will not simply be responded to but will be ripped to shreds at the hands of those who disagree or who may be smarter or more worldly than I. (They number in the billions.) Fortunately, that has not happened very often here, with some notable exceptions. What is hard to convey to new bloggers and publishers is that the debate is almost always civil, and that those naysayers who denigrate and tear down what they perceive as ignorance are not worth listening to. They are not teachers. I welcome disagreement, but I will tune out those who voice it with cynicism regardless of the validity of their response. When I read those constant smirkers, I wonder if they would treat younger learners the same way? Luckily, it seems, few of them are in classrooms.

Despite all of this, for me, right now, the rewards far outweigh the risks. I just wish I knew better how to convey that to those who see the scales tipping steeply in the other direction. And I wish I could help them understand that the angst I still feel every time I press "publish" is a good thing on balance, not something to avoid as much as to embrace as a path to a greater awareness of myself and of the world around me.

Source: tinyurl.com/9swana

The Wrong Conversations

School "reform," as it's currently being touted, isn't much of a reform at all. It's simply trying to figure out how to do better what we've been doing for 150 years. It misses the point. We don't need better any longer, we need different. Really different. I'm not sure what it will take to push this conversation to a point where real change occurs. But for now, it's up to each one of us to focus on what real changes we can effect in our own practice, in our own classrooms, and in our own schools.

28 Sep 2010 09:40 am

(This is gonna be a tough post to write. Not that blogging shouldn't be tough to begin with. But this one feels like it might be harder than most.)

By all accounts, it's been a crappy week for education. To be honest, I haven't participated in much of it, but reading the accounts from **Chris** (tinyurl .com/2ensxdh) and **Bud** (tinyurl.com/43rqgus) and others, and some of the Tweets from Sunday's **Education Nation** (tinyurl.com/2gxakmo) sessions, it's hard not to sense the anger, frustration, sadness and even paranoia that has infected our little part of the education world. While I know it was all heartfelt and sincere, I think I turned it all off on Sunday when a Twitter thread started to assume that certain books about the mess we're in had been somehow pulled from Amazon by NBC so as not to interfere with its one-sided reality about what fixes we all need to make education better. It goes without saying that it was much more fun watching Tucker win his soccer game and the Jets beat up on the Dolphins than watch the attempted dismemberment of the profession live and in Tweeting color.

But the last few days have me wondering a few things, among them, how many people are really tuned into this "conversation," how many of those will still be tuned in a month from now, and, the toughest one, are we just asking the wrong questions to begin with?

NBC understands as well as anyone the short attention span theater that is most effective to deliver a message to an increasingly dumbed down populace in this country. Crank up the machine for a few days of flooding, intensive market-ing under the guise of "conversation" in sound bites and then run to the next crisis. And the irony is that education really is failing if the vast majority of people go no further than to tune into Brian Williams or Oprah for an hour, receive the intended message, and then return to their lives thinking schools are broken and that billionaire-funded charters are the answer. Mission accomplished. (Of course the greater irony is that "student achievement" really has nothing to do with the critical thinking necessary to even attempt to navigate this morass of pseudo research and rock star opinion.) My sense is that very, very few people are "engaged" in these ideas, and most of them that are are angry. And rightfully so. NBC has the money and the bandwidth and the agenda in their pockets. "We" have a lot of passionate, kid-loving change agents who see the world a bit dif-ferently and are growing increasingly frustrated at our lack of a seat at the table.

But I guess I'm just wondering, do we even want a seat at that table? Are NBC and Oprah, and to a certain extent even the growing heroes in the movement like Diane Ravitch engaged in a debate that, at the end of the day, is going to be worth the time and energy we're spending on it?

And this is where it gets really hard for me, because while in my heart I know that to not fight these battles in the short term to preserve the very best of what schools and classrooms are and can be would dishonor the teachers and students currently in the system, I'm continually persuaded that at the end of the day, the focus on "fixing" schools occurs at the expense of a focus on expanding the learning opportunities we give our students. I wish the two were the same, that better learning was seen as the impetus for better schools. But right now, to the mainstream at least, better "knowing" means better schools. Say what you will about online social learning tools, the networks and communities that so many of us are engaging in do afford deep, rich learning in ways that physical space cannot match. (And yes, we can say the same about physical space.) The mainstream is not yet open to the opportunities for learning our students now have, due in large measure to these technologies, and it's nowhere near open to the idea that because of these innovations, the best outcome for our kids may be "schools" that look very little like what they look like today.

We need to be open to those ideas and more.

This post, **"We're Not Waiting for Superman, We Are Empowering Superheroes"** (tinyurl.com/34helx6) by **Diana Rhoten** (tinyurl.com/3lt2mur) of **Startl** (tinyurl.com/yengb9k) is the latest of many to push me in this direction. In it, she suggests that we are faced with a "massive, radical, design challenge," that "we need to reframe the problem and the conversation, from one about re-forming schooling to one about re-thinking education and re-imagining learning." So much of what she says in this post makes sense to me. Here's one snip especially:

> Our vision of technologically enabled learning is not one of the lone child sitting at her desktop (or laptop) passively consuming PDFs or browsing web pages. We believe the potential of technology for learning is much greater. We believe its power resides in its ability to deliver active and interactive experiences where a learner participates in the very construction of knowledge by crafting and curating, mixing and re-mixing information with digital tools, a process which can be and should be greatly augmented by online and offline social interactions between friends, in a community of peers, or an extended network of people (both professional and amateur) who share her interests.
>
> Technology is just a tool. Its effects ultimately depend on the people who use them, how and where. Thus, technology does not negate the role of people or place in learning, but it does change their definitions and their dynamics. And, so just as we design new technologies for learning, we must also consider the contexts for learning that will facilitate their best use . . . whether that is at school, at home, at the library, on the job, or a place we have not yet imagined.

And she frames what I think is a coherent (for these times, at least) vision for innovation on the edges (echoing **Christensen** [tinyurl.com/665wns2]) when she says:

> We believe the edge is the place in the system where the risk of failure and the opportunity for success are most allowable, and we want to be the people who take the risk to demonstrate the opportunity. We're not Pollyannaish about the challenges of working on the edge. We know much of what we try will fail; that's what innovation is about. We also know that it will take time for the work we support to travel from the early adopters to the mainstream, but we don't see an alternative. Better to demonstrate what could be than to wait for what might be.

Exactly. We should all be innovating, testing new models, failing, reflecting, trying anew, sharing the learning with others who are working on the edges in their own classrooms and projects. It's one of the great pieces of what we do at **PLP** (tinyurl.com/c48f92), because we are innovating and succeeding and failing and rethinking on the edge. And I know that's hard because it's not valued and supported in most places, and I know most teachers simply can't or won't. It's too hard. There's no time. Too many barriers. But those that can, must right now. Because the reality is we simply don't have the media, the money or the muscle to compete with the current narrative about schools, and to fret over that fact I think cuts deeply into what energy we do have to think clearly about what's best for our kids. And because in the long run, this conversation can't be about schools first. It has to be about learning. And through that lens, we need to be advocates for whatever is best for our kids, whether at times that might be a technology over a teacher, an online community over a school, a passion based project over a one-size fits all curriculum, a chance to create with strangers of all ages over a classroom of same-age kids working hard to game the system. Those types of innovations will at some point get the notice of the mainstream.

Let NBC and Bill Gates and Oprah have at the "fixing schools" conversation. Let's keep our energies and our laser like focus on the learning, in whatever form that takes.

Source: tinyurl.com/2wesd8w

Index

CORWIN
A SAGE Company

The Corwin logo—a raven striding across an open book—represents the union of courage and learning. Corwin is committed to improving education for all learners by publishing books and other professional development resources for those serving the field of PreK–12 education. By providing practical, hands-on materials, Corwin continues to carry out the promise of its motto: **"Helping Educators Do Their Work Better."**